HIGH-PERFORMANCE
FIBRE COMPOSITES

HIGH-PERFORMANCE FIBRE COMPOSITES

J. G. Morley

Wolfson Institute of Interfacial Technology,
University of Nottingham, U.K.

1987

ACADEMIC PRESS
Harcourt Brace Jovanovich, Publishers

London   Orlando   San Diego   New York
Austin   Boston   Sydney   Tokyo   Toronto

ACADEMIC PRESS LIMITED
24/28 Oval Road,
London NW1

United States Edition published by
ACADEMIC PRESS INC.
Orlando, Florida 32887

British Library Cataloguing in Publication Data

Morley, J.
  High performance fibre composites.
  1. Fibrous composites
  I. Title
  620.1'18   TA418.9.R4

  ISBN 0-12-506445-4

Printed in Northern Ireland at The Universities Press (Belfast) Ltd.

Preface

The aim of this book is to provide engineers, designers and materials scientists with some insight into the nature and potential of high-performance fibre composites. The characteristics of these materials are governed primarily by the properties of the reinforcing fibres, which have high strengths and high stiffness values and occupy a large proportion of the volume of the composite. Advanced composite materials are fabricated in ways that enable the advantageous characteristics of the fibres to be exploited as effectively as possible in an engineering structure. The book therefore deals both with reinforcing fibres and with the physical principles of reinforcement. To provide some measure of perspective, brief mention is made of the characteristics of traditional structural metals and of composites that utilize a small admixture of fibres to enhance the strength and toughness of a matrix.

The processes used in the manufacture of various reinforcing fibres are outlined and an account is given of the more important factors that govern their characteristics. Reinforcement is considered both from the point of view of the elastic characteristics of the composites and of the mechanics of failure. These issues are central to a basic understanding of advanced fibre composites.

Since the early 1960s a very considerable world-wide investment has been made in research and in the engineering applications of fibre composites. This is reflected in the large numbers of research papers and technical reports that have been published. A multiplicity of engineering applications has developed with specialized design requirements and material properties. Although this book deals only with the fundamental characteristics of high-performance composites, a fairly extensive list of references to research papers, review articles and books dealing with specialized aspects of the field is included. It is hoped that these will provide convenient points of departure for readers wishing to explore particular aspects of the subject in greater depth, but a fully comprehensive treatment of all aspects of high-performance composites has not been attempted. The reinforcing fibres described have been chosen primarily to illustrate fundamental principles and to give some insight into the various manufacturing routes that can be taken. The composite properties described are related to the physical processes occurring at the microstructural level and deal primarily with polymeric matrix systems. The major points of difference with metal matrix systems are discussed where appropriate.

The plan of the book is as follows. In Chapter 1 the general characteristics required of a structural material are outlined and an introduction given to the advantages of metals and fibrous composites in this context. Some of the processes used in the fabrication of fibrous composites are also outlined.

Chapter 2 deals with a wide range of reinforcing fibres including continuous fibres such as glass, boron, carbon, silicon carbide and aromatic polyamide fibres. Various types of discontinuous reinforcing fibres are also described.

Chapter 3 deals with the elastic properties of fibre composites. The relationships between the elastic constants of an isotropic material are first developed and methods of computing the various elastic moduli of fibre composites from the properties of their component parts are then outlined. The elastic properties of laminae and laminates are discussed at an elementary level; the mathematical development of the theory of anisotropic elasticity is kept to a minimum. Finally in this chapter the relative behaviours of continuous and discontinuous fibres are compared.

Chapter 4 deals with the development of matrix cracks in brittle matrix composites. Simple unidirectionally reinforced materials and also laminates are considered. Comparisons are made between theoretical predictions and experimental results.

Chapter 5 is concerned with the fracture of fibrous composites under tensile loading and deals primarily with unidirectionally reinforced systems. Various processes that inhibit failure, such as interfacial debonding and fibre "pull-out" are described. Composite failure is considered, both in terms of the statistics of fibre failure and from the viewpoint of the micromechanics of crack growth, and this is extended to cover hybrid systems.

Chapter 6 deals with fracture under more complex off-axis and compressive loading conditions. The behaviour of laminates is considered, including the effects of combined in-plane stresses.

Chapter 7 is concerned with some aspects of the long-term strength of fibrous composites. An account is given of the factors controlling the damage tolerance of laminates. The mechanisms of fatigue damage in unidirectional systems and laminates are also outlined. Finally, a brief mention is made of environmental effects on the mechanical properties of polymeric matrix composites.

I am grateful for assistance from many sources in the preparation of this book. Photographs of manufacturing processes and composite engineering components have been generously provided. I would like to thank all who have allowed me to reproduce photographs and diagrams from their published work and to their publishers for permission to use them. My understanding of composite materials has developed over many years during contact with many colleagues, initially in Rolls Royce Ltd. and subsequently in the University of Nottingham, and their contributions are much appreciated. I would like to thank the Wolfson Foundation and the Science Research Council for their generous support of my work in Nottingham.

It is apparent that a book of this nature must draw extensively on the published work of many authors. References are made in the text to specific contributions and these are warmly acknowledged.

J. G. MORLEY

Contents

Chapter 3
Elastic Properties of Fibrous Composites

Chapter 7
The Long-term Mechanical Properties of Composites in Engineering Applications

1
Outline of the General Characteristics of Structural Materials and the Fabrication and Applications of Fibre Composites

1.1 GENERAL REQUIREMENTS OF STRUCTURAL MATERIALS

A structural material is required to be strong and stiff enough to carry loads without being deformed significantly by them. For most applications a low-density material has obvious advantages. At the same time it should be easily shaped and joined to other components so as to form an engineering structure. It should be resistant to a range of corrosive environments and it should provide all these facilities as inexpensively as possible. For a considerable period of time metals provided the best compromise in meeting these requirements but, over the last few years, they have been supplanted for many applications by fibre-reinforced composite materials. There are now many types of composite materials in commercial use that show particular advantages in certain applications.

Fibre-reinforced composites also have their limitations. A simple composite consisting of unidirectionally aligned fibres in a matrix has highly anisotropic characteristics, since the fibres contribute to the strength and stiffness primarily in the direction of fibre alignment. For most engineering applications a structural material has to support loads applied in a variety of directions. It then becomes necessary to design a composite structure so that the optimum proportion of fibres is aligned in the necessary directions. Efficient composite structures can be manufactured, but correspondingly greater complexity is introduced into the design process. The elastic characteristics of fibre-reinforced composites can now be predicted with reasonable accuracy, but their various failure processes are not yet fully understood.

1.2   STRENGTH AND TOUGHNESS IN ENGINEERING MATERIALS

Strength and toughness are primary requirements in structural materials. The toughness of metals is achieved by a quite different process from that used in the design of fibre composites. Before outlining means by which the toughness and the resistance of a material to crack propagation are obtained, we first consider the theoretical strength of an ideal solid.

The tensile fracture of an ideal homogeneous elastic solid should occur by the pulling apart of adjacent planes of atoms. The force of attraction between the atomic planes varies with their distance of separation, being zero when the atoms occupy their equilibrium positions. For small deformations the force increases approximately linearly with increasing separation (Hooke's Law). The attractive force between the atoms must eventually reach a maximum value, corresponding to an applied stress of σ_{max}, and then decrease with increasing separation of the atomic planes (Orowan, 1949). This relationship can be represented approximately by a sign curve (see for example, Cottrell, 1964). Hence the external applied stress σ will be given (see Figure 1.1) by

$$\sigma = \sigma_{max} \sin{(2\pi x/\lambda)} \tag{1.1}$$

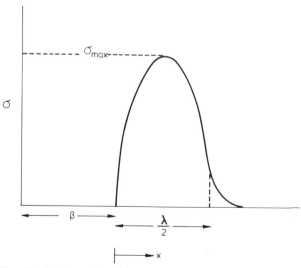

Figure 1.1. Tensile fracture of an ideal solid. (Redrawn from Cottrell, 1964.)

where x is the displacement from the equilibrium separation β and $\lambda/2$ represents the limiting displacement for interatomic cohesion. For small values of x,

$$\sigma = \sigma_{max} 2\pi x/\lambda \tag{1.2}$$

and from Hooke's Law,

$$\sigma = Ex/\beta \tag{1.3}$$

where E is the Young's modulus of the material. From equations (1.2) and (1.3):

$$\sigma_{max} = \lambda E/2\pi\beta \tag{1.4}$$

We now assume that the work done in separating two planes of atoms is equal to the surface energy γ of the newly created surfaces, so that

$$2\gamma = \int_0^{\lambda/2} \sigma_{max} \sin(2\pi/\lambda)\, dx = \sigma_{max}\lambda/\pi \tag{1.5}$$

From equations (1.4) and (1.5), we have

$$\sigma_{max} = (E\gamma/\beta)^{1/2} \tag{1.6}$$

Inserting appropriate values for λ and β in equation (1.6) gives

$$\sigma_{max} \simeq E/10 \quad \text{or} \quad \varepsilon_{max} \simeq 0.1$$

where ε_{max} is the maximum theoretical elastic strain shown by a solid at fracture.

These values are about an order of magnitude greater than present day strong elastic reinforcing fibres and about two orders of magnitude greater than conventional materials in bulk form. This discrepancy is due to the presence of cracks and flaws in practical materials. It is convenient to represent a crack as an elliptical cavity within an elastic plate (see Figure 1.2). The stresses round such a cavity have been calculated by various people, following the original work of Inglis (1913) (see e.g. Timoshenko and Goodier, 1951). The stress σ_y developed at the tip of the ellipse in the direction of loading when a unidirectional tensile stress σ is applied to the plate is given by

$$\sigma_y = \sigma(1 + 2a/b) = \sigma[1 + 2(a/\rho)^{1/2}] \tag{1.7}$$

where a and b are the semi-major and semi-minor axes of the ellipse and ρ is the radius of curvature of the ellipse at the tip of its major axis (crack tip). The applied stress is thus increased locally by a factor of three due to the presence of a circular hole. Also the local stress enhancement at the tip of a crack increases as the length of the crack increases and the radius of curvature of its tip decreases.

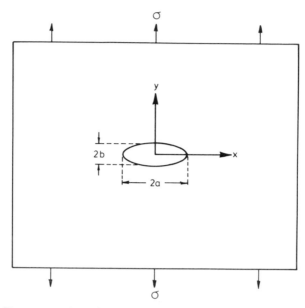

Figure 1.2. Representation of a crack by an elliptical cavity.

In addition to the stress σ_y acting in the direction of loading, a stress σ_x also acts perpendicular to the direction of loading. This is zero at the crack tip, increases to a maximum value of short distance from the crack tip and then falls progressively at increasing distances. Values of σ_y and σ_x calculated for an elliptical crack for which $a = 100b$ are shown in Figure 1.3. The maximum value of σ_x is about one-fifth of the maximum value of σ_y and this is reached at a distance from the crack tip roughly equal to the radius of curvature of the crack tip. It is apparent that stresses exceeding the theoretical strength of a solid can be generated by cracks of suitable geometry. When the material is homogeneous and elastic, the stress-raising ability of a crack is independent of its absolute size.

If a crack in a brittle material is to extend, energy must be supplied at least equal to surface energy of the newly created surfaces. This energy can be provided by the elastic relaxation of a stressed solid as the crack extends. Griffith (1920) showed that, for the situation illustrated in Figure 1.2, the amount of energy supplied by elastic relaxation is equivalent to the complete relaxation of an elliptical zone around the crack having an area equal to twice that of a circle whose diameter is the crack length. If the plate is assumed to have unit thickness the amount of energy released is therefore given by $2\pi a^2(\sigma^2/2E)$ and the rate of release of strain energy with increasing crack length is given by $d/da\,(\pi a^2\sigma^2/E)$. Since two surfaces are

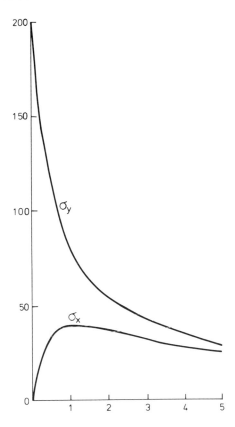

Figure 1.3. Local stress enhancement near the crack tip calculated for points on the x axis. Distances are expressed in terms of the radius of curvature of the crack tip. (Redrawn from Cook and Gordon, 1964.)

produced at each end of the crack, the minimum rate of absorption of energy is given by $d/da\,(4\gamma a)$, where γ is the surface energy. The crack becomes unstable when the rate of release of strain energy with increasing crack length equals the rate of energy absorption. Thus the critical condition for crack growth is given by

$$\pi a \sigma^2 / E = 2\gamma \tag{1.8}$$

so that

$$\sigma = (2\gamma E / \pi a)^{1/2} \tag{1.9}$$

Equation (1.9) applies to plane stress (thin sheet) conditions. For plane strain conditions the stress for unstable crack growth is given by

$$\sigma = \left(\frac{2E\gamma}{\pi(1 - v^2)a} \right)^{1/2} \qquad (1.10)$$

where v is the Poisson ratio of the material.

When a crack becomes unstable its rate of growth increases up to a limiting value which is about one-half of the velocity of transverse elastic waves in the material. In brittle materials, internal flaws, caused for example by foreign inclusions, can initiate fracture at points ahead to the main crack on planes different from that in which the main crack is propagating. Where the crack faces intersect, a step is formed that has a quasi-parabolic shape in the plane of the primary crack face. These effects produce essentially linear topographical features that radiate out from the point of initial fracture.

In the case of metals, plastic deformation occurs at the tip of a crack as a result of the high stresses developed there. When the zone of plastic deformation is small compared with the length of the crack, the bulk of the material behaves elastically and equation (1.9), suitably modified, can be applied. We now have to include the work done in deforming the material at the crack tip prior to fracture occurring there by plastic flow. In practice the surface energy term can be neglected, since the work done during plastic deformation is typically greater by several orders of magnitude. If we replace 2γ by G_c, the total resistance of the material to crack extension, we can write equation (1.9) as

$$\pi a \sigma^2 / E = G_c \qquad (1.11)$$

A crack may propagate by a crack opening (Mode I), forward shear (Mode II) or by parallel shear (Mode III); see Figure 1.4. It is convenient to substitute $K_{1c}^2 = G_{1c}E$ for the crack opening mode of failure (for plane stress conditions) so that $K_{1c} = \sigma(\pi a)^{1/2}$ and is termed the critical stress intensity factor. This relationship between K_{1c} and G_{1c} applies only when the crack length is small compared with the dimensions of the material and requires modification to deal with other stress conditions (see e.g. Paris and Sih 1965).

The large amount of work required to cause a metal to fracture by crack propagation makes metal alloys, in general, insensitive to damage during monotonic loading and is a basic reason for their widespread use as structural materials. Under cyclic loading, however, cracks can propagate incrementally at relatively low stress levels until they reach the critical size for unstable extension. This is termed fatigue failure and can be atmosphere-dependent. It is the source of most of the failures in service of metal structures and sets a primary limitation on the use of metals in highly

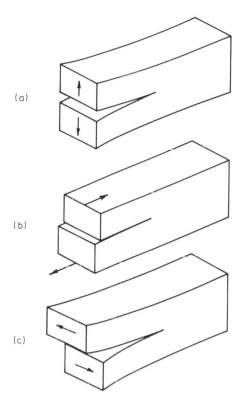

Figure 1.4. Crack extension modes. (a) (Mode I) crack opening mode; (b) (Model II), forward shearing mode; (c) (Mode III), parallel shearing mode.

stressed cyclically loaded structures. Other limitations of metals are due to their relatively high densities, which for practical alloys increase in proportion to the elastic modulus of the material. The ratio of elastic modulus to density is therefore sensibly constant for most metal alloys and this sets a limitation on the use of metals where movement (oscillatory, rotational or translational) is required of a structure.

Fibre-reinforced composites offer certain advantages over metals. This is because they employ a mechanism different from ductility to resist failure by the growth of a crack from a region of damage. Because ductility is not required of the load-bearing fibres they can be formed from brittle materials that have high elastic moduli and low densities. These materials, when in the form of thin fibres, can be made very strong in general, although weak regions will exist as a consequence of surface damage or the presence of flaws. Although failure may occur in the fibres at widely spaced flaws, these failures do not cause immediate failure of the composite structure. This is because the matrix in which the fibres are embedded transmits loads from the end of broken fibres to the still-intact fibres, so that the load-bearing ability of a fractured fibre is impaired only locally. Also, the structure of the composite inhibits the growth of cracks transversely to the fibres. This is because the interface between the fibres and the matrix presents a barrier against crack propagation into the fibres. When the matrix is brittle and has a lower failing strain than the fibres, as is the case for example with glass-fibre-reinforced polymers, the fibres bridge a matrix crack and hinder its propagation.

The characteristics of the interface are of considerable importance in controlling the toughness and resistance to crack propagation in fibre-reinforced composites. If the fibres are strongly bonded to the matrix, the composite will have brittle fracture characteristics. If the interfacial bond is very weak, the composite will have very low strengths when loaded transversely to the fibres and will also have low strengths when loaded in shear in the plane of the fibres. The strengths under these loading conditions, even for optimum fibre–matrix bonding, are always very much less than the strengths developed in the direction of the fibre alignment. This is a fundamental limitation of fibre composites but can be overcome to some extent by aligning fibres in various directions. When maximum structural efficiency is required, this is done by arranging for the composite to be formed from several layers, each containing unidirectionally aligned fibres. The fibres within each layer are generally aligned in a different direction from those in adjacent layers. In this way strength and stiffness can be provided in several directions, but with a concommitant reduction in specific properties compared with a unidirectional system. Nevertheless, the strength and stiffness values of fibre-reinforced laminates are superior to those of metals on a weight for weight basis. It is also possible to control the degree of anisotropy of a component by controlling the alignment, position and type of the fibres within it. Although this introduces a further degree of complexity into design and fabrication it enables the characteristics of a component to be matched to its required performance more closely than is possible using homogeneous materials. Fibre-reinforced polymer matrix laminates show a loss of strength under cyclic loading, but this is progressive and manifests itself initially by a reduction in the stiffness of the material. This is a much more favourable mode of failure than the growth of fatigue cracks in metals.

1.3 FABRICATION TECHNIQUES AND APPLICATIONS OF FIBRE-REINFORCED COMPOSITES

A comprehensive account of the various processes that can be used in the fabrication of composites and the engineering applications of these materials is beyond the scope of this book. In this section we confine ourselves to an outline of the processes used in the fabrication of some of the more widely used composite materials.

The fabrication of the composite structure and the engineering component is achieved in one operation. It is therefore necessary to consider both the properties of the materials and the geometrical form of structure to be manufactured in choosing the most appropriate fabrication process. Various types of fibres are available, the most widely used being glass, carbon, boron and "Kevlar"* (an aromatic polyamide). The matrix may be a polymer, a metal or a ceramic. In many cases it is convenient to combine the fibres and the matrix in an intermediate or precursor form that is then used for the manufacture of the engineering component.

1.3.1 Polymer Matrix Composites

Polymers are formed of very long convoluted chain-like molecules that may contain many thousands of repeated units. The backbone of the chain is formed primarily from carbon atoms; the other atoms (e.g. hydrogen and nitrogen) included in the chain are attached to the carbon backbone. The bonds between the carbon atoms forming the chain backbone are aligned at 109° with each other and the atoms can rotate about these bonds. As a consequence, the individual chains are very flexible and polymers are generally characterized by low elastic moduli and low strengths. They are generally quite brittle, particularly at low temperatures, and most are of no structural value at temperatures in excess of 200°C. Their surfaces are soft and fairly easily eroded and their mechanical properties can be degraded by water absorption Nonetheless, the convenience of fabrication they offer outweighs these disadvantages in many applications.

Polymeric solids can be divided into two groups—thermoplastic and thermosetting polymers. In thermoplastic polymers the interchain forces are weak. They can be overcome at fairly low temperatures, allowing relative movement between the chains to occur so that the material may be deformed or extruded. The polymer becomes rigid again when the temperature is reduced. In thermosetting polymers cross-links are formed by chemical bonding between the polymer chains to produce a rigid 3-dimensional network. The mechanical properties depend on the numbers and types of cross-links developed during curing. This usually involves a heating process but it can be carried out at room temperature. A final post-cure heat treatment at a relatively high temperature is given when it is important to minimize any long-term changes in the polymer structure that

---

* Kevlar is a trade name of DuPont.

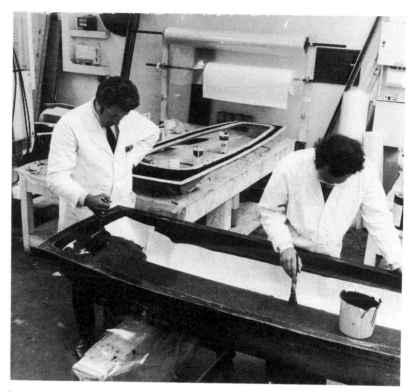

Figure 1.5. Hand lay-up process. (Courtesy of Pilkington Reinforcements Limited.)

might occur over a prolonged period in service. Thus, in contrast with thermoplastics, thermosetting polymers remain rigid at elevated temperatures and have to be fabricated to the final required shape of the component being manufactured before cross-linking occurs.

Various fabrication processes are available for use with polymeric matrix materials. The hand lay-up process is simple and convenient and makes use of an open mould. The fibres are used in the form of a woven cloth or as a mat of short fibres randomly oriented in a plane (chopped strand mat). The fibre sheets are placed in the mould by hand and the polymeric matrix added by painting or spraying (Figure 1.5). Additional sheets of fibres are then added and the process repeated until the required thickness is obtained. Room-temperature-curing epoxy or polyester thermosets are commonly used to form the matrix.

Alternatively, a bundle of continuous fibres (a roving) can be chopped into short lengths and sprayed on to the mould surface together with the

Figure 1.6. Spray-up process: deposition of resin and chopped rovings simultaneously into a mould. (Courtesy of Pilkington Reinforcements Limited.)

liquid matrix (Figure 1.6). As with the hand lay-up process, successive layers are compacted using rollers. The open-mould process used with either the spray-up or hand lay-up means of application requires little capital investment and is very versatile. When an ambient-temperature-curing polymer matrix is used, large structures can be fabricated and the method is used exensively for the production of small and medium-sized boats. The inner surface of the mould generates the outer surface of the boat, and is usually coated with a layer of pigmented resin before the reinforcing fibres and matrix are added.

The bag-mould technique makes it possible to use the essential features of the open-mould process together with control over pressure and temperature. A flexible sheet is used to compact the fibres and matrix and pressure is applied to the sheet in various ways (see Figure 1.7). Atmospheric pressure can be used, in which case a vacuum pump is used to evacuate the mould cavity. Alternatively, the flexible sheet can be pressurized so as to compact the material in the mould and appropriate curing temperatures can be obtained using an autoclave. This process can also be used for the manufacture of large components.

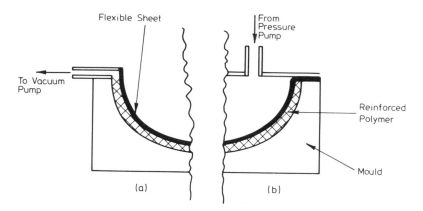

Figure 1.7. Bag moulding techniques: (*a*) vacuum bag; (*b*) pressure bag.

Figure 1.8. General view of filament winding process. (Courtesy of Pilkington Reinforcements Ltd.)

Tubular structures and pressure vessels can be manufactured by filament winding. In this process a rotating mandrel, having the internal dimensions of the component required, is rotated and overwound by a strand of fibres preimpregnated with a polymeric matrix in liquid form. Several layers of fibres are wound at predetermined angles and the matrix is partly or completely cured before the fabricated component is removed from the mandrel (Figure 1.8).

High-performance composite components, required to be manufactured to close geometrical tolerances from fibre layers aligned in various directions, can be fabricated using matched dies that are heated (Figure 1.9). Unidirectional fibre sheets, usually impreganted with an epoxy resin (prepreg), are used in their construction. The shapes of the various layers are chosen to correspond with the geometrical form of the component to be manufactured. The quantity of matrix present in the prepreg is closely controlled and is usually supplied partly cured and tacky, so that the various layers adhere to each other and maintain register in the mould before it is closed. Further curing then takes place in the mould when it is closed to form the component. The fabrication of high-performance polymer matrix composites has been discussed by Johnson (1983).

Sandwich structures, in which a lightweight honeycomb core is bonded between multiple laminated sheets, are widely used for aircraft and other lightweight applications.

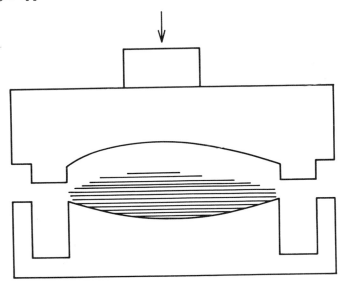

Figure 1.9. Schematic arrangement for matched die moulding, using material in the form of preimpregnated partly cured laminae.

(a)

(b)

Figure 1.10. Aero engine compressor; stator and rotor shown separately. Composite material components are manufactured in glass-fibre-reinforced epoxy novolac resin and glass-fibre-reinforced polyimide resin. (Courtesy of Rolls Royce plc.)

Unidirectionally reinforced fibre composite structures can be manufactured by the pultrusion technique, in which a continuous unidirectional array of fibres is impregnated with a liquid polymer and drawn though a heated die which removes any excess matrix material. The technique can produce a high volume fraction of fibres in the composite and has been used for the manufacture of rigid electrical insulators for high-voltage application. Partial or complete cure takes place as the material passes through the die that gives the cross-sectional shape of the material to be produced, rod or I-beam, for example. Fibre-reinforced polymers can also be fabricated using injection moulding processes. Woven cloth reinforcement can be placed in a mould that is then closed, and a polymeric matrix in liquid form is injected so as to infiltrate the fibre array. Alternatively, liquid polymer premixed with short fibres can be injected at high pressure into the cavity of a split mould and allowed to cure. An example of a complex engineering component (aero engine compressor) fabricated largely in fibre reinforced plastics is shown in Figure 1.10.

1.3.2 Metal Matrix Composites

Metals are crystalline solids. They can deform plastically because of the presence of mobile imperfections in the crystal lattice known as dislocations. The edge dislocation is the easiest to visualize and can be regarded as being produced by inserting an extra plane of atoms part way into the crystal structure (Figure 1.11). There is considerable distortion of the crystal structure close to the edge of the extra plane of atoms. Under the

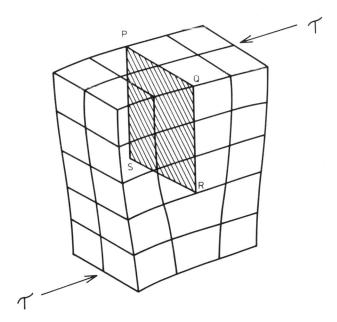

Figure 1.11. Schematic representation of an edge dislocation as an extra plane of atoms (PQRS) inserted into the crystal structure. The applied shear stress τ causes the dislocation to move through the crystal perpendicular to the plane (PQRS).

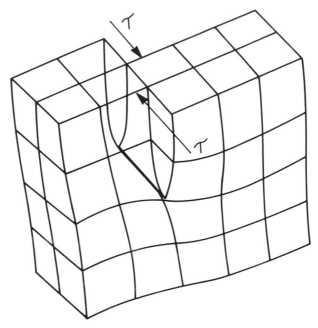

Figure 1.12. Schematic representation of a screw dislocation. Under the action of the applied shear stress τ the dislocation moves vertically downwards in the diagram.

application of a shear stress the region of distortion moves across the crystal, thus producing a shear deformation. The lack of directionality that is characteristic of metallic bonding permits dislocations to move at relatively low-shear stresses. This is in contrast with ceramic materials, in which the highly directional nature of the interatomic bonding prevents dislocation movement except at very high temperatures. Considerable deformation is produced in a metal when large numbers of moving dislocations are present. These allow sections of crystal to glide past each other on slip planes, generating slip steps on the surface of the material.

Edge dislocations allow deformation to occur in a direction perpendicular to the dislocation line. Screw dislocations (Figure 1.12) allow deformation to occur parallel to the dislocation line. Mixtures of edge and screw dislocations normally occur and new dislocations are generated as deformation proceeds. Mobile dislocations reduce the strength of a metal to values very much below the ideal tensile strength and the ideal shear strength.

However, local high stress concentrations associated with a crack or notch can be relieved by plastic flow, provided the crystal can deform in a sufficient number of directions. It can be shown that five independent glide systems are required operating at similar values of shear stress in order that each grain in a polycrystalline material can accommodate the deformation of the grains surrounding it. Since dislocations are mobile at low stresses in pure metals, it is necessary to restrict their movement in order to use homogeneous metals as structural materials. This can be done in various ways. Metals are used in polycrystalline form and their strength can be increased by decreasing the size of the individual grains. Dislocation movement can also be impeded by the inclusion of foreign particles within the grains. This may be achieved by producing solid solutions or by precipitating small particles from a supersaturated solution using an appropriate thermal treatment.

Some of the processes used in the manufacture of metal matrix composites are similar to those used for the fabrication of polymer matrix systems.

Boron-fibre-reinforced aluminium is produced by first manufacturing unidirectional tape in which a layer of boron fibres is bonded to a tape of aluminium alloy foil (Kreider and Prewo, 1974). The fibres can be bonded to the foil using a polymeric adhesive that volatilizes at the temperatures used to fabricate the composite (the fugitive binder technique); alternatively, the boron fibres can be bonded to the aluminium foil with a spray of molten aluminium particles delivered from a plasma torch. The particles cool rapidly, thus avoiding the chemical reactions that take place between boron and aluminium at elevated temperatures. Composites are fabricated by placing layers of tape in a mould and applying heat and pressure to deform and consolidate the aluminium around the fibres. Temperatures, times and pressures need to be carefully controlled in order to minimize chemical and mechanical damage to the strong brittle fibres. It is also necessary to prevent oxidation of the boron fibres and the aluminium during hot pressing. Because of the high cost of boron fibres, the use of this material is limited to aircraft structures. An alternative approach, in which the individual fibres are coated with a uniform layer of metal by freeze-coating, has been shown to be feasible for the manufacture of silica-fibre-reinforced aluminium. The cold fibre is passed at high speed through a molten bead of metal formed at the end of a narrow tube fed from a container held at constant temperature (Arridge and Haywood, 1967). The metal in contact with the fibre is cooled very rapidly, thus inhibiting adverse chemical reactions as the aluminium solidifies and is drawn away to form a uniform coating. Hot pressing techniques, similar to those used for boron fibre aluminium composites, have been used to consolidate the layers.

Metal matrix composites can also be formed by infiltrating with molten metal a bundle of fibres held within a mould. Usually the mould is evacuated so as to avoid cavity formation in the composite. This technique is useful when the matrix wets the fibres and where no serious adverse reactions are encountered between the molten matrix and the fibres.

Metal matrix composites can also be manufactured by forming the reinforcing fibres *in situ* through the controlled solidification of a molten alloy. In this way reinforcing rods or platelets are precipitated from the matrix, their alignment being controlled by the direction in which the material is traversed through a temperature gradient. The reinforcing elements are usually formed from intermetallics or carbides and are strong, brittle and temperature-resistant. The process is more easily controlled if eutectic compositions are used; suitable growth conditions are more difficult to maintain for off-eutectic compositions. Attention has been focused on high-melting-temperature alloys based on nickel, chromium, niobium and cobalt among others. The component can be formed to the required final shape by casting and the heat treatment necessary to produce the required phase separation can be carried out subsequently with the component supported in a ceramic mould (Thompson and Lemkey, 1974).

1.3.3 Ceramic Matrix Composites

These materials fall into two groups. Fibre-reinforced cements and plasters have been used in the construction industry on a large scale for a considerable time. Horse hair has been used to reinforce plaster and steel rods to reinforce concrete. In these applications the matrix is initially plastic and hardens at ambient temperatures as a consequence of hydration reactions. The second class of ceramic matrix systems is designed for use at high temperatures. Although the compressive strengths of ceramics are relatively high, these materials are weak in tension because flaws are inevitable in practical materials and the works to fracture of ceramics are low. (Very high strengths can be attained if flaws are eliminated.) The fibres are used to inhibit the growth of cracks when tensile loads are applied, thus enhancing the tensile strength of the material without degrading their compressive strengths.

Glass-fibre-reinforced cememt (GRC) is in widespread use in the building and construction industry. The fibres considerably increase the tensile flexural and impact strength of the cement matrix (Blackman *et al.*, 1977). Typical applications include cladding panels, machine casings, litter bins, cable connections boxes and pipe-work. The glass fibres used are resistant to the alkaline environment of the cement. Various compositions can be used to suit the requirements of different applications, but typically GRC contains about 5% by weight of glass fibres in a Portland cement/sand mortar. Portland cement consists of a mixture of calcium oxide, aluminium oxide and silica. When water is added, interlocking adhering crystals of hydrated calcium aluminosilicates are produced that bond the material together to produce a rigid solid. This is a slow process and periods of up to 28 days might have to be allowed to elapse in an atmosphere of 100% humidity to achieve optimum properties.

GRC components are usually produced by a spray method. Simultaneous sprays of cement/sand mortar paste and short lengths (12 mm to 50 mm) of glass fibre bundles are deposited from a hand-held or mechanized spray gun. A uniform felt of fibres and mortar is deposited on to the mould surface to build up the required thickness of GRC. The fibre bundles are aligned randomly in the plane of the sheet. Hand-held rollers are used to remove any entrapped air and to ensure that the material conforms with the mould surface. The typical daily output from one spray gun is about 2.5 tonnes. Flat sheets and more complex sections can be produced in a continuous form.

A limited amount of GRC is produced by first mixing fibres and matrix in a container to form a slurry which is then cast in an open mould. Press moulding is used for small components.

The improved tensile strength of glass-fibre-reinforced cement over that of the unreinforced matrix is particularly useful during the initial transportation and handling of GRC structures. Over a period of time the contribution of the fibres is reduced by chemical and physical interactions, causing some reduction in the tensile and impact strength of the composite material, but the long-term strengths are adequate for non-structural and quasi-structural applications. The material has not been used so far as a primary load-bearing material.

The strength of carbon can be enhanced by the incorporation of carbon fibres. This material is now in widespread use in aircraft brakes because of its very high thermal capacity and low density. Carbon-fibre-reinforced carbon composites can be manufactured in various ways. The fibres may be infiltrated with a polymeric material that is then carbonized; alternatively, pitch may be used. Several impregnation and carbonisation cycles may be required to achieve the required level of porosity and very high heat treatment temperatures are needed to convert the matrix into graphite crystals. The fibre array may also be infiltrated with a carbon-containing gas and then heated to decompose the gas and deposit carbon on the fibres. Again, many impregnation cycles are required to produce a composite having low porosity.

2
Reinforcing Fibres

In this chapter the manufacturing processes and the mechanical properties of various types of reinforcing fibres are discussed. Some fibres, e.g. glass fibres, may be cut into relatively short lengths for convenience of handling for use in certain types of fabrication techniques. However, in advanced fibrous composites, the aspect ratio (length-to-diameter ratio) of the fibres is always high. This is necessary in order to utilize effectively the high strength and stiffness of the reinforcing fibres when tensile loads are applied to the composite material in the direction of fibre alignment. Under these loading conditions the matrix serves primarily to transfer external loads to the fibres and to distribute the loads between fibres. The mechanical properties of the matrix are of greater importance under other conditions of loading and these issues are discussed in Chapter 6.

The mechanical properties of importance in reinforcing fibres are high strength, high elastic modulus and low density. Fibres of commercial significance must also be convenient to manufacture and hence inexpensive. They must also be available in a suitable form for subsequent composite fabrication. Unidirectional fibre arrays are appropriate for some engineering applications, but in most cases it is necessary to align the fibres in various directions. The fibres therefore have to be produced in a form that lends itself to these requirements. The bulk of reinforced composites in current use have polymeric materials as a matrix, but ceramic matrix and metal matrix systems have also been developed. This chapter deals with fibres used in composites that are constructed from separate component parts, i.e. fibres and matrix.

2.1 GLASS FIBRES

2.1.1 Factors Affecting the Tensile Strength of Glass

Strengths approaching the theoretical ideal strength of a solid can only be observed if stress-concentrating flaws are eliminated and all deformation processes other than elastic ones are suppressed. Silica-based glasses have characteristics that correspond quite closely with those of an ideal elastic solid. However, it should be noted that the surface of glass is observed to yield in compression when high local pressures are applied (Marsh, 1964).

It is generally considered that the structure of glass can be described as a random 3-dimensional network lacking periodicity, but with an energy content comparable with that of the corresponding crystalline structure (Zachariasen, 1932). Hence, the coordination number for a glass-forming atom must be similar both in the glassy and crystalline states, and the following simple rules apply when the oxide of an element A is capable of existing in the glassy state.

(1)   An oxygen atom is linked to not more than two atoms of A.

(2)   The number of oxygen atoms surrounding A must be small (3 or 4).

(3)   The oxygen polyhedra share corners with each other, not edges or faces, and form 3-dimensional networks.

(4)   At least three corners of each oxygen polyhedron must be shared.

The regular structure of a crystal is distorted in the case of a glass to give a random network that lacks periodicity, although the individual structural units are identical in both cases. The random network theory has been extended by the suggestion that "holes" bounded by oxygen atoms necessarily exist in the glass network. In the case of multicomponent glasses these "holes" form excellent positions for the cations of the non-network-forming component. The presence of foreign cations in such "holes" in the network is in agreement with the requirement that they should not cause a great increase in the potential energy of the system. The random network theory is supported by x-ray diffraction spectra. It can be deduced from the absence of small-angle scattering that the bonding of silica glass must be essentially continuous. Interatomic distances are found to be very similar to those of the crystal cristobalite, indicating that the structure is based on silicon–oxygen tetrahedra.

It follows from the above that the essential components of the mineral glasses are mutually soluble oxides with a valence 3 or more. Much the most common of these is silica, although glasses based on P_2O_5 and GeO_2 can be manufactured. Other oxides are commonly introduced to reduce the manufacturing temperature, to reduce the tendency for recrystallization at intermediate temperatures, and to improve the optical properties or chemical stability of the glass. Silica can be prepared as a glass without other oxide additions, although high manufacturing temperatures are required.

The highest strength values for solids in bulk form have been observed for fused silica and most of the studies of the factors affecting strength of silicate systems have been carried out on this material. The earliest strength estimates were made by Griffith (1920), who heated a portion of silica rod 5 mm in diameter to a high temperature in an oxy-hydrogen flame and drew down the hot viscous material to form a central thinner section about 1 mm in diameter. After this was allowed to cool it was observed that the drawn down portion could be deformed elastically to a radius of 4 or 5 mm without breaking, the bending loads being applied through the massive but relatively weak ends of the specimen, with the material showing almost perfect elasticity. Hillig (1962) made quantitative observations on test samples manufactured by similar techniques. Measurements were made at low temperatures by allowing liquid nitrogen to flow over the samples. Elastic strain values were observed considerably in excess of the value of about 0.1 that would have been expected from a flaw-free brittle solid. Moreover, the elastic modulus of the samples was observed to increase with increasing strain, contrary to the behaviour expected from an ideal elastic solid.

The techniques of sample preparation used by Griffith and Hillig were developed further by Morley et al. (1964) to produce specimens that could easily be tested in tension. Fibres were prepared from silica rods about 1 mm in diameter. The ends of the rods were heated to a high temperature so as to form small spherical end beads which could be held in simple grips for tensile loading. The centre section of the rod was then heated to a very high temperature using a gas flame and drawn down in a controlled manner to form a fibre about 25 μm in diameter. There was no difficulty in applying the very high tensile stresses necessary to break the fibre through the relatively massive ends of these specimens, despite the damage caused to the material at the point of attachment.

Silica fibres produced in this way are observed to have tensile strengths of about 6 GN m^{-2} when measured at room temperature in air. The fibres show appreciable variability in tensile strengths, the variability in their measured strength values being considerably greater than that which would have been expected from possible experimental errors. Investigations into a possible correlation between flame conditions, rod size, the position of the rod in flame, the duration of the initial heating period and other factors gave no positive result. No correlation was obtained between fibre strengths and any other conditions of manufacturing that could be investigated in a systematic manner. Morley et al. (1964) concluded that the source of the

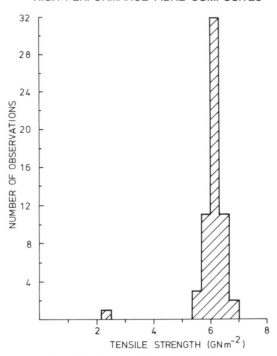

Figure 2.1. Tensile strength of silica fibres, tested in air at room temperature. (Redrawn from Proctor *et al.,* 1967.)

variability in fibre strengths either lay in some aspect of the manufacturing conditions that was not under control, or else was due to some inherent variability in the properties of the silica itself. These results were confirmed and extended by Proctor *et al.* (1967), who used the same techniques to manufacture their experimental fibres. Although they took great care in handling the fibres under test, in order to avoid as far as possible any surface mechanical damage, the observed variability of fibre strengths was still quite appreciable (Figure 2.1). The individual observations were thought to be accurate to ±3%, so that the scatter represents a genuine variability in fibre strengths.

The fibre strengths are observed to be extremely dependent on surface perfection: the slightest mechanical contact with any other solid causes severe loss in strength. Etching with hydrofluoric acid also leads to considerable reductions in strength. This is rather surprising, since the same treatment enhances the strength of silicate glass specimens with damaged surfaces (Proctor, 1962). Apart from these factors, the high-strength surface is remarkably stable, being unaffected by immersions in water and aqueous solutions of common reagents and by length of storage in a normal atmosphere. However, fibre strengths are very susceptible to testing conditions and extensive studies have been carried out on the effect on fibre strength of temperature, the atmospheric environment of the fibre when under load, and the rate of application of the load.

The strength of silica when under stress has long been known to be dependent on the surrounding atmosphere; moisture is a particularly important factor. Surface water is removed *in vacuo* and fibre strengths are found to be very much higher under these conditions, a state of equilibrium being reached after about 30 min. This change in strength is reversible, a normal strength distribution being observed when the fibres are again exposed to water vapour or normal atmospheric conditions. Strength varies with the water vapour content of the atmosphere, reaching values similar to observations *in vacuo* at moisture contents of about 10 p.p.m. The effect of various gaseous environments, including nitrogen and argon, have been investigated, but the fibre strengths are found to be sensitive only to water vapour.

From −269°C up to 200°C the effect of temperature on the strength of silica fibres is reversible. Above 200°C irreversible effects take place. The reversible effects are summarized in Figure 2.2; the results are characterized by a very rapid rise in fibre strengths as the temperature is decreased below 0°C. These effects can be explained in terms of a reaction between silica under stress and water vapour (Charles, 1958). This links the effects of varying temperature on both the reaction rate between silica under stress and water and also on the mobility and concentration of water on the silica surface. The still-appreciable variability in fibre strengths implies a distribution of very small flaws and the reaction with water vapour is assumed to occur at the tips of these cracks, where very high stresses must exist. The net effect of the chemical reaction is to increase the severity of the flaw, thus reducing the observed strength of the material. The constancy of strengths *in vacuo* over the range −80°C to 200°C can be accounted for by the increase in reaction rate and mobility being compensated by the reduction in availability of absorbed water. In air, the shape of the curve indicates that this compensation is not achieved until 0°C and the increase in strength up to 200°C can be explained by assuming that the dominant factor is loss of water from the surface as the temperature increases.

The region of virtually constant strength *in vacuo* from −80°C to 200°C, together with the region of approximately constant strength from −269°C to −196°C and the increase of strength in air from 0°C to 200°C, suggest that direct thermal energy effects in the molecular structure do not significantly influence the experimental strength of the material. The experimental evidence points very strongly to the presence of very small stress-concentrating flaws that control the strength at these levels.

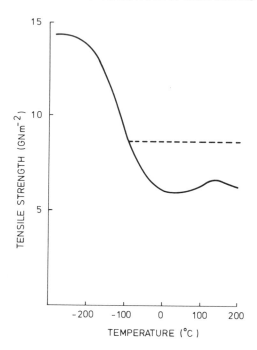

Figure 2.2. Average strengths of silica fibres as a function of temperature. Results for fibres tested *in vacuo* shown (– – – –). (Redrawn from Proctor *et al.,* 1967.)

Above 200°C silica fibres suffer permanent loss of strength. The strengths fall both as the temperature and period of heating before testing are increased and this behaviour is summarized in Figure 2.3. The strength values shown refer to fibres stressed to failure at the temperatures indicated. If the fibres are cooled to room temperature before testing, the strengths are further reduced. The failure is caused by local surface stress-raising flaws which seem to be caused by local fluxing of the silica owing to contamination with dust.

When strong silica fibres are loaded to a stress level that is comparable with, but less than, the stress level necessary to cause fracture under normal tensile test conditions, it is found that failure takes place after a time. The time interval before failure increases as the applied stress decreases and the effect is known as static fatigue.

Results of constant tensile stress static fatigue experiments are shown in Figure 2.4. The applied stress is plotted against the logarithm of the time to fracture and the slope of the plot can be regarded as a measure of the static fatigue effect. There is virtually no static fatigue at −196°C and only a slight

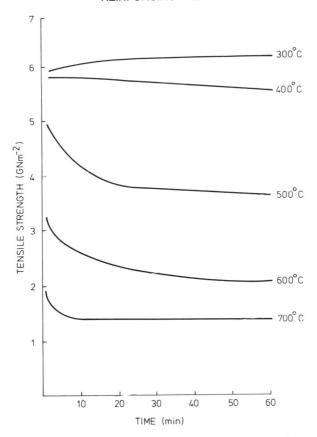

Figure 2.3. Average strengths of silica fibres tested at elevated temperatures in air. (Redrawn from Morley, 1964.)

effect at room temperature *in vacuo*. At room temperature in air there is a marked variation in "time to fracture" with applied stress but delayed failures are not observed below a stress level of about $2.75 \, \text{GN m}^{-2}$. The effects of atmospheric and temperature changes shown in Figure 2.4 can be explained in terms of a stress-induced moisture interaction proceeding at a rate that depends on both stress level and moisture concentration. Thus, lowering the temperature, or testing *in vacuo*, delays fracture by reducing mobility, interaction rates and concentration of the surface water. On this argument, the applied stress limit at about $2.75 \, \text{GN m}^{-2}$ represents the minimum value needed to activate the water–silica reaction at the tips of the stress-concentrating flaws at room temperature in air.

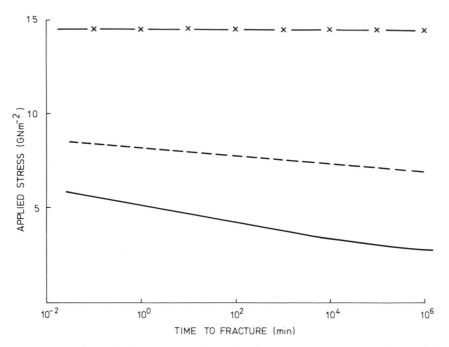

Figure 2.4. Static fatigue of silica fibres in air at room temperature (———); *in vacuo* at −196° (——×——); *in vacuo* at room temperature (− − − −). Curves indicate average values. (Redrawn from Proctor *et al.*, 1967.)

Above the static fatigue stress limit there is a roughly linear relationship between applied stress and log (time to failure).

Thus,

$$(\sigma - \sigma_0) = (1/n) \ln K - (1/n) \ln t \tag{2.1}$$

where σ is the applied stress, σ_0 is the fatigue limit, t is the time to failure and n and K are constants.

From the line drawn through the results at room temperature in air in Figure 2.4, values for n and K are obtained; σ_0 is equated with the experimental value 2.75 GN m^{-2}.

Equation (2.1) may be obtained from the more general expression

$$\int e^{n(\sigma - \sigma_0)} \, dt = K \tag{2.2}$$

when applied stress σ is held constant ($\sigma > \sigma_0$). Equation (2.2) suggests the concept of a total "stress life", which could be applied to strength measurements as normally obtained under constantly increasing stress. At a constant rate of increase ($\sigma = qt$) the breaking stress from equation (2.2) is given by,

$$\sigma_b = \sigma_0(1/n) \ln{(nqK + 1)} \qquad (2.3)$$

On substituting values for n, K and σ_0 into equation (2.3), strengths at much higher loading rates can be predicted. Proctor *et al.* (1967) measured strengths of silica fibres for "times to fracture" from less than 10^{-2} s up to almost 10^2 s. The different rates of loading have a significant effect on the measured strength values, which are predicted to an accuracy within about 10% by equation (2.3).

The static fatigue strengths of various types of silicate glass fibres are affected by the presence of water vapour in a similar manner to those of silica fibres. Much greater reductions in strength are observed when fibres are loaded in tension in the presence of mineral acids (Hogg, 1983). The effect depends on the acid type and concentration and also on the glass composition. Aveston and Sillwood (1982) observed that the long-term strength of E glass fibres was reduced considerably in the presence of dilute sulphuric acid, while "Cemfil"* glass fibres—originally developed as an alkali-resistant fibre (see §2.1.7)—appeared to be unaffected.

The high and consistent tensile strengths observed with silica fibres at liquid nitrogen temperatures have been utilized by Mallinder and Proctor (1964) to measure the effect of large elastic strains (~ 0.12) on the elastic properties of this material. Elastic strains of this magnitude are comparable with the limits of interatomic cohesion, so that experimental confirmation of the predictions of equation (1.6) would be expected to be obtained by this means. In particular, the reduction in the slope of the elastic stress–strain curve as the tensile stress approaches σ_{max} would be expected to be observable (see §1.2). However, the elastic characteristics of fused silica are very different from those that would be expected from an ideal elastic solid.

The elastic modulus of the material, measured for small elastic strains, is about one-fifth of the value that would be expected from the known force constants for the interatomic bonds and their number per unit cross-section. The value of Young's modulus, measured as the slope of the stress–strain curve, is about 70 GN m^{-2} for small strains and increases with increasing strain in contrast with the behaviour expected from simple theory. It has been suggested that this is due to the very "open" molecular structure of silica, which allows deformation to occur by changes in bond angle as well as by bond stretching.

It should be noted that flaws do not seriously impair the reinforcing efficiency of glass or other fibres providing they are widely spaced. Under

---

* Cemfil is a trade name of Pilkington Bros. plc.

these conditions fracture produces discontinuous fibres that have high length-to-diameter ratios and hence reinforcement efficiencies. This issue is discussed at greater length in §§3.11 and 5.1.

2.1.2 General Outline of the Fibre Manufacturing Process

The feasibility of drawing heat-softened glass into fine fibres was known to glassmakers in antiquity and indeed is older than the technique of glass blowing. However, it was not until early in the 1930s that satisfactory processes were developed for the commercial manufacture of glass fibres. Glass fibres were first used as a reinforcement for polymers during the 1940s. The scale and range of their application have increased progressively since that time and they are now used on a considerably greater scale than any other type of reinforcing fibre. Their manufacture can best be considered as taking place in the following three stages.

(1) Glass manufacture.
(2) The conversion of molten glass into glass fibres, i.e. fibre drawing.
(3) The conversion of glass fibres into materials suitable for the fabrication of reinforced composites.

The technology of manufacturing continuous glass fibres has been discussed in some detail by Loewenstein (1973) and Lowrie (1967). Glass is manufactured from suitable raw materials such as sand, limestone and boric acid which are weighed and mixed together in appropriate proportions and melted in a glass-making furnace. The glass produced in this way may be fed directly into the fibre drawing furnaces called "bushings". This is termed the direct-melt process. Alternatively, the glass may be made into marbles that are annealed and cooled to room temperature and then stored for future use. They are then fed into a fibre drawing bushing and reheated. This process is more convenient in some circumstances than the direct-melt process.

The fibres are produced by drawing away at high speed a solidified filament of glass from a molten drop exuding from a nozzle located on the underside of the bushing, which is an electrically heated platinum-alloy container. The number of holes in the base of a bushing is usually 204 or a multiple thereof. Molten glass flows through each nozzle, the meniscus formed at the point of fibre formation being controlled by the surface tension of the glass, its viscosity and the rate of fibre drawing (Figure 2.5). The glass fibre is wound on a cylinder called a "collet" rotating at high speed and located below the bushing. The collet is usually covered during the winding process by a removable paper or plastic tube onto which the strand of fibres is wound. The individual fibres being drawn from the nozzles in the bushing first pass through a light water spray and then over

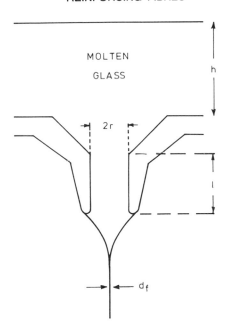

Figure 2.5. Illustrating the flow of glass through a nozzle in the base plate of a bushing. (Redrawn from Loewenstein, 1983.)

an applicator that transfers a protective and lubricating size onto the filaments. This is necessary to avoid excessive damage to the fibre surfaces during subsequent processing. By this means, high fibre strengths can be preserved. The fibres are then gathered into a bundle of filaments called a strand, which passes through an oscillating traverse that moves the strand rapidly across the collet. In this way the strand is distributed uniformly across the length of the collet, successive portions of the strand being aligned at small angles to each other to facilitate subsequent unwinding. A small water jet may be arranged to play on the traverse in order to prevent the fibre size drying on it. The overall manufacturing process is indicated diagramatically in Figure 2.6. The diameter of the collet is usually between 200 mm and 300 mm and the collet rotates at speeds of up to about 7000 rpm, giving maximum fibre drawing speeds of the order of $100 \, \text{m s}^{-1}$. The fibre drawing process is usually intermittent since the process has to be stopped when the thickness of the fibre layer or "cake" on the collet reaches about 25 mm. This is necessary because of the requirement to remove from the cake the water applied with the lubricating size . There are also limits to the increases in cake diameter that can be tolerated while maintaining the rate of fibre drawing within limits acceptable to the fibre drawing process.

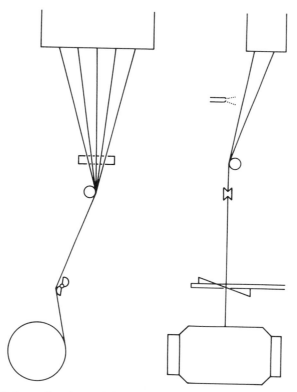

Figure 2.6. Schematic view of fibre manufacturing process. (Redrawn from Loewenstein, 1983.)

The fibre drawing process is sometimes made continuous by arranging for the fibre strand to be transferred to a second collet by an indexing mechanism that allows a fully wound cake to be removed from one collet while the strand is being wound on a second collet. To facilitate removal of the cake, whether or not fully automatic winding processes are used, the collet is fitted with longitudinal segments that expand radially by a limited amount under centrifugal force, thus gripping the plastic or paper tube and the glass fibre cake. The collet is designed to have a low inertia to facilitate rapid acceleration and deceleration.

2.1.3 The Fibre Drawing Process

As outlined above, glass fibres are made by the rapid attenuation of drops of molten glass flowing through nozzles under gravity. The glass flows through the nozzle forming a drop; when the drop becomes large enough it falls away leaving a fibre attached to the molten material flowing through the nozzle, providing the viscosity of the molten glass is within the range suitable for fibre drawing. The thickness of the fibre depends on the rate at which filaments are wound on the collet, but the quantity of glass turned into fibre is entirely a function of the rate of flow of glass through the nozzles and is independent of the rate of attenuation. The rate of flow of molten glass through a nozzle can be described generally by the Poiseuille equation:

$$F = Ar^4h/l\eta \qquad\qquad (2.4)$$

where F is the rate of flow, A is a constant, r is the diameter of the nozzle bore in its narrowest cyclindrical section (usually in the range 1–2.5 mm), l is the length of the cylindrical section of the nozzle (usually in the range 2–6 mm), η is the viscosity of the glass, and h the head of glass.

As the fibre is drawn away from the nozzle, the drop of glass takes the form shown in Figure 2.5. The process depends on the balance between the effects of surface tension and viscosity. For fibre drawing to be successful the viscosity of the glass has to be within a fairly narrow range (50–100 Pa s) at the drawing temperature. At lower viscosities the glass is too fluid and falls away from the nozzles as drops. At higher viscosities the tension in the fibre during attenuation becomes too high and the rate of flow of glass through the nozzles becomes too low to maintain the drawing process. The function of the bushing, therefore, is to provide a plate containing several hundred nozzles all at a uniform temperature and to ensure that the glass is also at a uniform temperature so that the fibres being drawn from all the nozzles are of a uniform diameter. It is also necessary, of course, that the glass should have other suitable characteristics, since factors other than viscosity are of importance. These are discussed in §2.1.4.

Owing to cooling effects, the viscosity of the glass increases as it flows through the nozzle, so that it is not possible exactly to predict the rate of flow of glass from nozzles of different designs, since the temperature gradients and hence the viscosity gradients may differ. In the case of bushings fed with glass from marbles, convection currents arising from the melting of the cold glass at the top of the bushing may also introduce further effects. An increase of the level of the glass can produce a substantial layer of glass of high viscosity at the top of the bushing that can reduce the flow of glass through the nozzles despite the increased static head of glass. The bushings are usually constructed from platinum 10% rhodium or platinum 20% rhodium alloys, the latter being more resistant to distortion at elevated temperatures. Since a 408-nozzle bushing weighs approximately 2–4 kg, depending on whether it is a direct-melt or marbles bushing, and since the rate of production of such a bushing is only of the order of 100 tonnes of material per year, it follows that the cost of manufacture is affected significantly by the required capital investment in platinum alloys.

There are some practical limits to the degree of attenuation that can be achieved during the drawing process. The stability of the meniscus is in part a function of the draw-down ratio, i.e. $2r/d_f$, where d_f is the fibre diameter (Figure 2.5). It is desirable to pass as much glass as possible through a nozzle within a given period of time, but the fibre drawing process becomes more difficult to control the greater the degree of attenuation at the meniscus. Ideally, therefore, the bushing should operate with the smallest possible nozzle diameters that give the desired output rates at the maximum possible operating temperatures. The extent to which this can be done is increased by cooling the surface of the molten glass as quickly as possible when it is being drawn from the nozzle. This is achieved using water-cooled shields placed between individual nozzles.

It is found to be more satisfactory to have a smaller bore and shorter cylindrical section than a larger bore and longer cylindrical section. When the fibres are being drawn at high speed, problems may occur where they come into contact with guides and other surfaces (see Figure 2.6). The stability of fibre sizes may be insufficient to protect the fibre surfaces from damage under these conditions, so that fibre strengths are reduced and drawing becomes impracticable.

2.1.4 Glass Characteristics and Manufacture

For the manufacture of glass fibres to be efficient, the composition of the glass must be very homogeneous and there must be a complete absence of impurity particles, since solid inclusions of even sub-micrometre dimensions act as stress concentrators that reduce fibre strengths to very low levels. Such inclusions could be specks of refractory material dislodged from a side-wall of a furnace, owing to variations in the level of molten glass, or metal dust particles rubbed off the feeding tubes of marble-fed bushings. A single fractured filament may cause adjacent filaments to fracture, thus leading to an unintentional interruption in fibre manufacture.

Inhomogeneity, arising from inadequately mixed raw materials or their inadequate dispersal during glass melting, can cause sudden changes in viscosity. These may adversely influence the fibre making process by changing the rate of flow of glass through the nozzles or by taking the viscosity of the glass outside the limits acceptable by the drawing process. On the other hand, gas bubbles in the molten glass, unless very big, do not cause filament breakage. Instead they become attenuated, forming capillaries several metres long in the glass filaments. The fibre drawing process can operate successfully over only a limited range of fibre viscosities and values of surface tension. Surface tension is not strongly dependent on composition and temperature of the glass, but the reverse is true of viscosity. Ideally, the correct viscosity of the glass should be obtained at the lowest possible operating temperature, since this prolongs the life of the platinum alloy bushings. The choice of operating conditions is constrained by the tendency of glass to recrystallize or devitrify over a temperature range extending below the liquidus temperature of the glass. At the liquidus temperature the rate of growth of the primary crystalline phase (the phase having the highest liquidus temperature) is zero. Above the liquidus temperature of the primary phase all crystalline material dissolves and its rate of solution increases with increasing temperature. As the temperature is decreased progressively below the liquidus temperature the rate of growth of the primary phase at first increases but then reaches a maximum and falls to zero as the temperature is further decreased. Other forms of crystalline material forming secondary and further phases show similar growth characteristics at lower temperatures. Crystal growth rates are strongly linked with glass viscosity (Morley, 1965). It is apparent that during the fibre manufacturing process the glass must pass through a temperature regime within which recrystallization, in principle, can occur.

It is important to prevent recrystallization occurring both during the formation of the fibre and earlier during the flow of the glass to the nozzle, since small particles of recrystallized material will drastically reduce the fibre strength. The liquidus temperature of the primary phase falls as additional components, not present in the primary phase, are added to the glass composition. Generally, this will also reduce the growth rate of the primary phase at temperatures below its liquidus temperature. On the other hand, some compositional changes can promote higher growth rates of the secondary phase, or additional phases, and the recrystallization behaviour of a complex glass composition is not analytically predictable. Hence improved glass compositions follow only as a consequence of continued development, and compositions found suitable for fibre manufacture tend to be retained. A very high proportion of all the continuous glass fibre produced is of the so-called E type. E glass was originally developed for electrical applications, but today these form only a small portion of the total market. Various other glass compositions have been developed to meet various requirements (see Lowrie, 1967). S glass fibres are a relatively recent development and have average tensile strengths of about 5 GN m^{-2} compared with 3.5 GN m^{-2} for E glass fibres. The approximate proportions of the main components of E glass and S glass are shown in Table 2.1.

Because of the rapid cooling the fibres experience during manufacture, the lower-density liquid structure of the glass is largely preserved in the fibres. This causes the fibres' elastic moduli to be rather lower than would be observed in an annealed glass of the same composition. The density and refractive index of the fibres are also lower than those of annealed glass. If the fibres are reheated, thermal compaction occurs and their physical properties approach those of annealed glass as the heat treatment tempera-

TABLE 2.1
Approximate compositions of various glass fibres
(main components)

Glass type	SiO_2 (%)	Al_2O_3 (%)	CaO (%)	MgO (%)	B_2O_3 (%)
E	55	15	19	3.0	7.0
S	65	25	—	10.0	—

ture is increased. However, the tensile strength of the fibres is reduced in a manner similar to that observed with fused silica fibres (see §2.1.1). In the case of E glass fibres this effect is apparent at temperatures as low as 150°C (Thomas, 1960).

Melting of glass is usually carried out in a tank furnace. This consists of a chamber constructed from suitable refractory materials in which the upper part serves as a firing chamber and the lower part contains the molten glass. For glass fibre manufacture the length of the tank furnace is several times its width and burners utilizing oil or gas are positioned opposite each other along each of the longer walls.

The glass is heated throughout its depth by a process of infra-red absorption and re-radiation. At high temperatures the effective thermal conductivity due to this mechanism can be very high but depends on the infrared absorption characteristics of the glass at melting temperatures. The raw materials for glass manufacture are fed continuously from one end of the tank furnace and molten glass flows out from the other end either directly to bushings, for direct fibre manufacture, or to be formed into marbles (usually 19 mm in diameter) for subsequent remelting for fibre manufacture.

Glass fibres can be manufactured in the forms required by the various manufacturing techniques for polymeric matrix composites described in §1.3.1. The term "strand" is given to a unidirectional bundle of fibres drawn from a single bushing. A roving is a bundle of essentially parallel strands wound up without twist to form a package. Fibre strands and rovings can be chopped into short lengths inexpensively to form a non-woven fabric in which the fibres are orientated randomly in a plane. Textile fabrics used for reinforcement utilize continuous strands or rovings.

2.1.5 Fibre Sizes and Coupling Agents

Fibre sizes and coupling agents are substances applied to the surface of the fibres, firstly to prevent excessive damage and loss of fibre strength during manufacture, and secondly to improve the bond between the fibre and polymeric matrix. The latter requirement is particularly important if the composite has to operate in the presence of water. The first lubricating sizes used contained starch and corn oil but, although effective in protecting the surface of the filaments during fibre drawing and subsequent textile operations, these substances prevent effective bonding between the fibres and a polymer matrix. Hence it is necessary to remove them, using solvents and heat treatment, prior to the application of coupling agents to promote bonding. If the mechanical treatment to which the fibres are subjected during manufacture is not severe (e.g. in non-woven products) so-called compatible sizes can be used. These offer some protection against mechanical damage to the fibre surface during manufacture and also promote bonding between the fibres and various types of polymeric matrix. The chemistry of coupling agents is complex and has been dealt with extensively for various types of reinforcing fibres by Plueddmann (1974). The issues will be given here only in outline.

When exposed to a wet environment a polymeric matrix material absorbs water. Although some of the water is present in the form of bound molecules, most is present as free water. This can occur as droplets within air bubbles and as aqueous solutions of impurities within resin cracks. Its presence causes crack growth in the matrix during freezing as well as loss of adhesion at the fibre–matrix interface (Nicholas and Ashbee, 1978). The latter effect is particularly apparent at temperatures above ambient. Various organosilane materials are used to improve adhesion. A number of theories have been put forward to account for the behaviour of coupling agents. The chemical bonding theory is the most important of these, bonding being assumed to take place through bifunctional compounds capable of interacting with the polymeric matrix and with certain sites on the glass surface. Other arguments are based on beneficial modifications to the matrix properties in the vicinity of the fibres brought about by the presence of the coupling agents. The surface finishes have to function despite the initial presence of water on the fibre surface and react with a thermosetting matrix resin during its cure. These substances are effective in preventing a deterioration of composite properties by degradation of the interfacial adhesion between fibres and matrix even after prolonged treatment in water at elevated temperatures.

2.1.6 Glass Fibres for Cement Reinforcement

Glass fibres used for the reinforcement of cement have to withstand the highly alkaline environment of the cement matrix. The strengths of conventional glass fibres are rapidly degraded under these conditions and special glass compositions are required for this application. The fibre manufacturing process is basically the same as described in §2.1.2, so that the glass compositions used have to meet the physical requirements of the fibre drawing process as well as preserving sufficient alkali resistance.

It has long been established that the addition of zirconia to sodium silicate glass systems increases their resistance to alkali attack, and it has been demonstrated more recently (Majumdar, 1970) that a similar chemical resistance can be developed in zirconia-containing glass fibres. Further investigation (Proctor and Yale, 1980) resulted in the development of glasses, basically in system $Na_2O-CaO-ZrO_2-SiO_2$, which possessed sufficient alkali resistance and also the other characteristics necessary for large-scale manufacture. Recently the alkali resistance of these fibres has been improved by a proprietary treatment to the fibre surface.

The requirement that the strength of glass fibre should not be impaired over long periods in the alkaline environment of a cement matrix is not simply associated with the resistance to chemical attack. For example, glass strengths can be increased by etching in hydrofluoric acid (Proctor, 1962). There is not necessarily a direct correlation between the rate of degradation of fibre strength in a liquid environment and that occurring under the same chemical conditions but with a different physical environment. Fibre strengths are related to the topography of the corrosion process and the physical and chemical environment in which it is occurring. As a consequence of these uncertainties in interpretation of the experimental data, tests designed to simulate as nearly as possible the conditions the fibres would experience in use have been devised. Also, since glass-fibre-reinforced cement products are required to retain their design characteristics over considerable periods of time, it is necessary to devise accelerated ageing tests. Such tests and their correlation with field tests carried out over periods of years under a wide range of climatic conditions are described by Litherland et al. (1981).

2.2 BORON FIBRES

2.2.1 Basic Manufacturing Process

Boron in filamentary form was the first of the new reinforcing fibres to be produced in any quantity. It has a high elastic modulus ($\sim$400 GN m^{-2}), a high average tensile strength (3.5 GN m^{-2}) and a low denisty (2.6 $\times$ 10 kg m^{-3}). It has been used to reinforce both polymeric materials and metals, particularly aluminium. The early technology of boron fibre manufacture has been reviewed by Wawner (1967). The mechanical properties of the first laboratory samples to be produced were described by

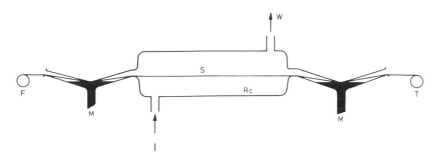

Figure 2.7. Boron fibre manufacture (schematic). M, mercury contact seal; S, substrate wire; F, feed spool; T, take up spool; R.C., reaction chamber; I, inlet gases; W, waste gases.

Tally (1959) as having a tensile strength of $1.6\,GN\,m^{-2}$ and maximum surface tensile strength in flexure of $2.4\,GN\,m^{-2}$. Boron fibres are normally produced by the deposition of boron from a mixture of hydrogen and BCl_3 on to an incandescent tungsten wire, but other techniques can be used. A review of various manufacturing processes has been given by Carlsson (1979).

The manufacturing process for the production of continuous boron fibres from a boron halide is shown schematically in Figure 2.7. The substrate tungsten wire, with a typical diameter of $12.7\,\mu m$ (0.5×10^{-3} in.), is heated electrically to the required temperature using a direct current and drawn through the reaction chamber at a suitable speed. Mercury seals maintain the appropriate conditions in the reaction vessel and also serve as electrical contacts. To increase production rates, several reactors are arranged in series. The wire passes through an initial cleaning chamber where it is heated electrically in the presence of hydrogen to remove surface contaminants before it enters the deposition chambers.

The deposition process can be described by the equation $2BX_3 + 3H_2 \rightarrow 2B + 6HX$, where X represents either Cl, Br or I. Successful deposition of boron from the BCl_3–H_2 system can be carried out over a composition range of 10–$60\,mol\,\%$ BCl_3. The rate of deposition and the nature of the boron coating formed is critically dependent on the substrate temperature. For temperatures below about 1000°C the rate of deposition of boron is very low. For high deposition temperatures above about 1300°C large crystallites can be formed and the fibres have low tensile strengths. At temperatures of about 1100°C deposition rates are sufficiently high to permit short residence times in the reaction chambers and the fibres produced have high tensile strengths. Fibres are usually manufactured with overall diameters of about

100 µm, but much larger diameter fibres (up to 236 µm) have been produced.

Galasso *et al.* (1966) found that boron deposited at a substrate temperature less than 1400°C was amorphous in form; it was tetragonal between 1400°C and 1500°C, and above 1500°C it was rhombohedral. The structure of amorphous boron appears to be built up in successive layers with a very large number of stacking faults between layers (Otte and Lipsitt, 1966). This type of structure, which may be described as either amorphous or microcrystalline, consistently yields a higher value of filament strength. The structure of the material deposited is dependent on the presence of reaction products in the reactant gas atmosphere as well as on the temperature of the substrate fibre.

A thin coating (~2 µm) of silicon carbide is sometimes added to improve the resistance to oxidation of the fibres at the expense of a reduction of about 10% in average fibre tensile strengths. The silicon carbide coating also protects the boron from reacting with aluminium at elevated temperatures and hence facilitates the fabrication of composites utilizing aluminium alloys as a matrix.

During manufacture a temperature gradient is developed along the fibre owing to the presence of the boron coating and the changing resistance of the substrate wire as it reacts with the boron coating to produce various tungsten borides. The existence of a temperature gradient makes it more difficult to maintain optimum deposition temperatures and various steps are taken to minimize these effects. The use of a multistage system is helpful in this respect, since the temperature gradient in each reactor can be small. Temperature gradients can also be minimized by the control of the gas flow and gas mixture. Deposition rates can be increased by preheating the reactant gases and by adding small amounts of various halides and fluorides that act as catalysts.

Hot spots can sometimes occur in the neighbourhood of the mercury seals and crystalline boron developed in these regions considerably reduces the strength of the fibre. Jacob *et al.* (1973) describe a radiofrequency-heated reactor by use of which mercury seals can be avoided and a uniform temperature can be maintained along the fibre during boron deposition.

2.2.2 Microstructure of Boron Fibres Grown on a Tungsten Substrate

Boron fibres are not homogeneous but consist of fairly complex assemblies of tungsten boride crystals in the core of the fibre surrounded by a thick coating of amorphous boron.

The core increases in diameter as the tungsten reacts with the boron

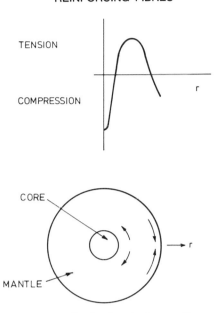

Figure 2.8. Illustrating stress distribution between the core and surface of a boron fibre. (Redrawn from Vega-Boggio and Vingsbo, 1977.)

during the growth of the boron coating (Adler and Hammond, 1969). As a consequence, compressive stresses are developed in the core and dilatational tensile stresses in the surrounding boron coating. Boron also possesses anelastic properties at elevated temperatures, deformations generated under load being recoverable during further heating when not under load. The rapid quenching of the outer layers of the filament at the end of the manufacturing process produces a compressive stress in surface layers of the material when the filament has cooled to a uniform temperature. The magnitudes of these stresses are of course a function of the details of the manufacturing process, but a typical stress distribution for 100 μm diameter fibres is shown in Figure 2.8. Since the fibre surface is placed in compression, it is apparent that the strengths of boron fibres will be less susceptible to surface damage than are those of glass fibres. However, radial cracks occur within the central regions of the fibres. They may extend axially over considerable distances but their radial extension is terminated by the compressive zones near the fibre surface.

The boron deposit grows in the form of cones that orginate at the substrate surface. These cones extend to the surface of the fibre and form nodules that give a characteristic corn-cob appearance to it. The nature of the substrate surface influences the appearance of the surface of the boron fibre as a result of the preferential nucleation of growth cones on singularities on the substrate. In particular, the regular axial alignment of the growth cones can be ascribed to parallel die markings on the surface of the tungsten wire substrate. Boron nodules are observed to nucleate preferentially on the die mark ridges produced during the manufacture of the tungsten wire (Vega-Boggio et al., 1977). Boron is lost from the surface layer by diffusion into the substrate to form tungsten borides at a rate that decreases with the increasing thickness of the boride layer. The overall effect is that the critical surface concentration for the nucleation of a boron nodule is reached earlier on a ridge than in a groove.

2.2.3 Fibre Fracture–Tungsten Halide Manufacturing Process

One would expect flaws to be generated at the core of boron fibres owing to the volume changes taking place as a consequence of the reaction between boron and tungsten. The effect of such flaws in reducing the strength of the fibre depends upon the nature and scale of the stress distribution developed in the fibre, on the fibre size and on the loading conditions. Vega-Boggio et al. (1977) have observed flaws, existing as small voids, within the boron mantle and quite near to the interface between the mantle and the core. They suggest that these voids are produced as adjacent rows of boron nodules (which have nucleated on parallel ridges on the tungsten wire core) coalesce with each other to produce a small cavity that extends an appreciable distance along the axis of the filament. Fibres that fracture because of such voids can be identified from the fractography of the fibre surface. The strength of such fibres correlates well with the flaw geometry and surface energy of boron if the voids are assumed to operate as Griffith flaws. (§1.2). It is further suggested (Vega-Boggio and Vingsbo, 1977) that flaws of this type can initiate the growth of the frequently observed radial cracks, since such flaws can be located in the region of hoop tensile stress present in the filament. The tensile strength of a boron filament would not be expected to be influenced significantly by a radial flaw extending only in a plane parallel to the fibre axis. However, fibre failure is observed to occur from the radial extremity of such a flaw and sometimes from the small stress-concentrating cavity that probably initiated the growth of the radial flaw in the first instance. Since the radial cracks are not observed to reach the surface of the fibre, it seems that they must be formed after the surface compressive stresses are developed at the completion of the manufacturing process.

Radial cracks reduce the transverse tensile strength of the fibres to very low values and these characteristics control the transverse strength of composites utilizing a high-strength metal matrix. Such fibres are readily split longitudinally, the two halves showing appreciable curvature and illustrating the presence of the large residual biaxial compressive stresses at the surface of the fibres as manufactured. The transverse tensile strength of a fibre can be conveniently measured by applying a diametral compressive stress (Kreider and Prewo, 1972). It is observed that the transverse strength is a function of the fibre diameter and the manufacturing process, as these factors affect the residual stress distribution within the fibre. The transverse tensile strength of the fibre measured in this way is very variable, since it depends upon the orientation of the radial cracks with respect to the loading direction. Standard fibres having a diameter of about 100 μm and produced by deposition on to a resistance heated tungsten wire show a sharply skewed transverse tensile strength distribution with a mode at about 200 MN m^{-2}. (This contrasts with the axial tensile strength of such fibres, which is usually in the range 2.7–3.5 GN m^{-2}.) Fibres of the same diameter produced by radiofrequency heating show a transverse tensile strength distribution that is much more nearly Gaussian with a mode increased to about 2.8 GN m^{-2}. Similar increases in fibre transverse tensile strengths are observed with fibres having larger diameter (140 μm). The average tensile strength of these fibres remains high (average 3.2 GN m^{-2}) even when they are subsequently coated with an outer 2 μm thick chemically protective layer of silicon carbide (Prewo and Kreider, 1972).

In addition to the flaws developed at the core owing to reaction between the boron and the tungsten substrate, discussed above, fibre failure has also been observed to occur from notches present in the initial core surface. The boron nodules growing from the substrate produce an irregular corn-cob fibre surface and stress concentrations are produced by this irregular geometry. Occasionally, crystalline or amorphous nodules protruding abnormally from the surface are present and these act as more severe stress concentrators. Foreign material and interlayer flaws developed during multistage manufacturing processes can be controlled by careful attention to cleanliness during manufacture and their presence is rare. The effect of surface flaws can be reduced by chemical polishing using nitric acid and the effect of this is comparable with that of hydrofluoric acid etching on surface flaws in the case of silicate glasses. The most serious flaws are eliminated first and the fibre strength approaches a constant value as the material is removed from the surface. If the surface compressive layer is removed by etching, the compressive stress on the core is increased. Hence the strength of fibres, which fail as a consequence of flaws within the core, is increased. The stress distribution is also modified by subsequent heat treatment and fibre strengths in excess of 5.5 GN m^{-2} have been observed by combining these treatments (Carlsson, 1979). Fibre strengths are mainly controlled by stress-concentrating flaws located in the central region of the fibre and the intrinsic strength of the amorphous boron is considerably greater than measured fibre tensile strengths. Boron fibres are much stronger in flexure than in tension, since flaws associated with the core are placed near the neutral excess and have less influence on flexure strength. Maximum surface tensile stresses in excess of $13\,GN\,m^{-2}$ have been observed for fibres tested in flexure and tensile strengths very considerably in excess of those observed for the complete filament can be measured for fragments of amorphous boron removed from the fibre (Wawner and Satterfield, 1967).

2.2.4 Summary of the Mechanical Properties of Boron Fibres

As with other brittle solids containing a distribution of stress-concentrating flaws, the tensile strength of boron fibres is variable and is dependent on the length of the sample tested. For a gauge length of 25 mm average fibre tensile strengths of about $4.0\,GN\,m^{-2}$ are achievable and standard production material has an average strength of about $3.4\,GN\,m^{-2}$. The Young's modulus of the fibre is typically $400\,GN\,m^{-2}$. The fibre density of tungsten wire substrate fibres is a function of fibre diameter, being influenced considerably by the proportion of the tungsten present. The density of standard fibres 100 μm in diameter is about $2.65 \times 10^3\,kg\,m^{-3}$. Thus the fibres have specific strength values about the same as high-strength glass fibres, but their specific stiffness values are about five times greater.

Boron fibres tested at room temperature after heat treatment show a progressive decrease in strength with increasing time. At about 300°C the strength falls to about 60% of its original value after an exposure period of one hour. When heated in air under an applied stress, the fibre strength falls rapidly to zero at a temperature of about 600°C (Veltri and Galasso, 1968). Fibres coated with a thin layer of silicon carbide resist oxidation. About 50% of the fibre strength is retained after 1 h heat treatment in air at 700°C. The strengths of boron fibres under compressive loading are at least as great as their tensile strengths. Also, the large diameters of the individual filaments allow these strengths to be developed in a composite structure, since failure in a buckling mode due to the presence of small voids in the matrix is less likely to occur.

Boron fibres and silicon-carbide-coated boron fibres show anelastic creep when subjected to stress at elevated temperatures (Prewo, 1974). The amount of deformation generated is dependent on the temperature and the applied stress and it has been suggested as being of value in the fabrication of engineering components of complex shape. The deformation is preserved after cooling to room temperature but is recovered if the fibres are again heated in the absence of an applied stress. The tensile strengths of

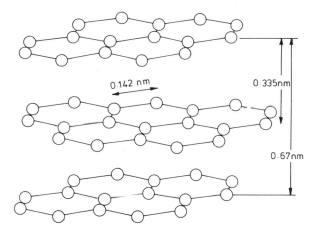

0.142 nm

0.335nm

0.67nm

Figure 2.9. Graphite lattice. (From Watt 1970.)

boron fibres coated with SiC are reported not to be affected by this treatment.

2.3 CARBON FIBRES—AN INTRODUCTION

The usefulness of carbon reinforcing fibres rests on the characteristics of the graphite crystal, illustrated in Figure 2.9. This is a hexagonal layer structure with the carbon atoms in the layers covalently bonded and closely packed having a separation of 0.142 nm. The spacing between the layers is relatively large ($\sim$0.335 nm) and the layers are bonded together only by van der Waals forces. The layer planes are stacked in an ABAB sequence, the unit cell having a height of 0.67 nm and an edge dimension of 0.246 nm. The elastic modulus measured parallel to the layer planes is 1015 GN m^{-2} and 35 GN m^{-2} perpendicular to the layer planes (Watt, 1970). Because of the highly anisotropic nature of the graphite crystal it is necessary to arrange for the layer planes to be aligned preferentially with the fibre axis to produce a carbon fibre having a high elastic modulus.

Elastic modulus values approaching those of a graphite single crystal can be attained with carbon fibres using commercially feasible manufacturing processes. Most of the types of carbon fibres now being manufactured are produced by the pyrolitic conversion of textile precursors. Carbon fibres produced in this way are not composed of perfect crystalline material and consist, in the main, of turbostratic graphite. In this material the atomic spacing within the layers is identical to that occurring in the graphite single

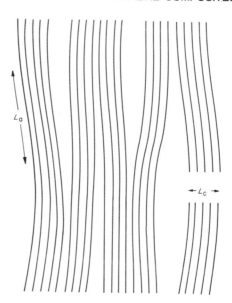

Figure 2.10. Schematic representation of the "microfibrillar" structure of high-elastic-modulus carbon fibres formed from graphite layer planes. The average length of straight sections of ribbons is given by L_a and the average thickness by L_c. (After Goodhew *et al.*, 1975.)

crystal, but the layers have a random stacking sequence with neighbouring layers being rotated out of register with each other. This rotational disorder increases the spacing of the layers to about 0.344 nm.

It is now generally accepted that the structure of carbon fibres produced from textile precursors can best be described as consisting of an interwoven array of "ribbons" of turbostratic graphite material. The "ribbons" are aligned preferentially with the fibre axis but there is appreciable "porosity" in the fibre as a consequence of this configuration (Figure 2.10). The axial elastic modulus of the fibre is increased the more nearly the crystal structure approaches a graphitic form and also as the orientation of the "ribbons" more nearly approaches that of the fibre axis.

Textile fibres based on polyacrylonitrile (PAN) are the most widely used precursor fibres, but high-elastic-modulus carbon fibres can be produced from cellulose precursors and also from pitch, which is first extruded into a fibre form. These are discussed in §§2.5 and 2.6. The fabrication and properties of carbon fibres and the suggested chemical processes occurring during conversion have been the subject of extensive reviews (see e.g. Goodhew *et al.*, 1975).

The choice of a precursor polymer fibre is limited by a number of requirements. It is necessary for the unwanted ingredients to be removed from the fibre during pyrolitic conversion without the fibre melting in the process. Indeed, where a continuous manufacturing process is used it is desirable for the fibre to retain a high tensile strength during all stages of the conversion process. Polymers can be classified into those that char, thus preserving the morphology of the precursor (as in the conversion of wood to charcoal), and those that coke, in which the morphology is not maintained (as in the conversion of coal to coke). However, most polymers do not fall completely into one or other of these categories and the classification is affected by the details of the pyrolysis process used. The conversion process involves the transport of material from the centre of the fibre to the exterior and, in addition, during the early stages of conversion, transport of oxygen to the centre of the fibre is usually required. Hence, there are advantages in using small-diameter precursor fibres. Exothermic reactions occurring during conversion are more easily controlled with small-diameter fibres. Further, as indicated above, the initial polymeric material must be converted almost entirely to carbon that is basically graphitic in form, although crystallographic perfection is not necessary. At the same time, the basal graphitic layer planes must be aligned preferentially with the fibre axis. This is achieved in various ways. For commercial reasons it is clearly desirable for the precursor material to be as inexpensive as possible and for a high proportion of the carbon present in it to be retained in the carbon fibre that is eventually produced.

A considerable degree of preferential alignment of the molecular chains of polymer fibres with the fibre axis can be obtained as a consequence of the manufacturing processes used (Watt, 1970). The carbon–carbon backbone of the polymer chains can be regarded as forming a "template" from which a graphitic structure is formed. In cellulosic (rayon) based precursor fibres considerable molecular disorientation occurs during pyrolysis and reorientation of the basal planes of the graphitic structure with the fibre axis is achieved by the plastic stretching of the carbon fibres at very high temperatures. By this means fibres having an elastic modulus of about $550\,\mathrm{GN\,m^{-2}}$ can be produced (Bacon, 1973). The technical difficulties associated with this high temperature deformation coupled with the relatively low carbon yield ($\sim 25\,\mathrm{wt\%}$) make this process less attractive commercially than the PAN-based process despite the lower cost of the cellulosic precursor.

In the case of PAN-based carbon fibres the manufacturing process depends on the degree of alignment achieved of the polymer chains in the precursor material. This alignment is preserved during subsequent pyrolysis by suitable processing conditions and carbon fibres having a high elastic modulus can be produced without the need for hot stretching. The relatively high yield of carbon, which can be in excess of 50% of the original carbon content of the polymer, reduces the importance of the higher cost of PAN compared with cellulose.

For successful commercial production it is found necessary first to stabilize the PAN textile fibre, both thermally and chemically, by carrying out an oxidative heat treatment in air at temperatures of about 200°C. This may involve stepwise heating over a temperature range. The fibres would contract on heating if unrestrained, thus reducing the preferential alignment of the polymer chains, and this is prevented during the oxidative treatment. Careful attention to the design and control of this stage is required if high fibre throughput is to be obtained without the danger of a runaway exothermic reaction occurring. The time required for the oxidative treatment varies according to the material being processed and the process arrangements, and may extend from less than 1 h (Clarke and Bailey, 1974) up to several hours (Watt and Johnson, 1970). The treated fibre is then heated to temperatures up to about 1200°C in a nominally inert atmosphere. Mechanical restraint of the fibre is not mandatory during this stage, the end-product being a fibre consisting almost wholly of carbon and having elastic modulus of about 200 GN m^{-2}. This elastic modulus is adequate for most engineering purposes. A third very-high-temperature (2500°C) "graphitizing" stage may be added to increase the fibre elastic modulus to about 3j0 GN m^{-2}. Further increases in fibre elastic modulus can be obtained through additional hot stretching procedures as used with cellulose precursors. The chemical and physical processes considered to take place during the conversion of the precursor PAN fibre through these processing stages are outlined in §2.4.

Pitch precursor fibres can be produced by a melt-spinning process analogous to those used for the manufacture of glass fibres and certain types of polymeric fibres. If the pitch is isotropic the fibres produced are isotropic and have low elastic modulus values even after carbonization to very high temperatures. High values of elastic modulus can be generated in these fibres by hot stretching at temperatures approaching 3000°C but it is more convenient to preferentially orientate the molecular structure of the pitch precursor fibre during the spinning operation. In order to achieve this it is first necessary to convert ordinary pitch into a mesophase (Bacon, 1979). Commercial pitches are heated to temperatures in excess of about 350°C, when dehydrogenative condensation reactions take place, the molecules aggregating to form extended sheets. This phase separates from the surrounding isotropic pitch in the form of small droplets. These coalesce and grow in size as the reaction continues. Eventually the anisotropic material becomes the continuous phase.

During spinning, the convergent flow at the approach to an orifice produces a longitudinal velocity gradient that tends to orientate the planes of the molecules with the fibre axis. Other velocity gradients occurring in the spinning process can orientate the molecular layers so as to produce either radial, circumferential or random alignments of the molecular layers in the fibre. The alignment of the layer planes with the fibre axis is retained, however. The anisotropic fibres produced in this way are stabilized and made infusible by oxidation at temperatures below the softening point of the pitch. They are then carbonized at temperatures in the region of 2000°C to convert the pitch to preferentially orientated graphitic material. It is not necessary to maintain a tensile load on the fibres during processing and a higher carbon yield (~80%) than those achievable by the conversion of polymeric precursors is attained.

Glassy carbon fibres, which have a very open arrangement of the graphitic "ribbons", can be produced from phenolic resins. These have low densities and because there is little or no preferential alignment of the layer planes with the fibre axis they have fairly low elastic moduli (Goodhew *et al.*, 1975). Because of the various precursor materials and processing conditions that can be used, carbon fibres are available in a very wide range of mechanical properties.

Carbon fibres were first manufactured on an appreciable scale in the 1950s, by the conversion of rayon into fibrous carbon material for thermal insulation, filtration and adsorption. Carbon fibre cloth and felt, pyrolysed to high temperatures, has been used since about 1960 to reinforce phenolic resins to form ablative materials for rocket and missile components. These fibres had relatively low elastic moduli and strength values, but there was rapid technical progress during the 1960s. High-elastic-modulus carbon fibre yarns ($\sim 170\,\mathrm{GN\,m^{-2}}$), produced by hot stretching cellulose based carbon fibres, were marketed during 1965 by the Union Carbide Corporation. Towards the end of the same year, carbon fibres having elastic moduli of approximately $350\,\mathrm{GN\,m^{-2}}$ were produced by Rolls Royce Limited in sufficient quantities to permit preliminary development work on engineering components. These fibres were produced by the thermal conversion of PAN precursor fibres under non-oxidizing conditions (see §2.4.1). Resin-impregnated warp sheet utilizing high-elastic-modulus carbon fibres ($\sim 200\,\mathrm{GN\,m^{-2}}$) based on a PAN precursor, and using an initial oxidative step to stabilize the fibres—a procedure previously investigated under laboratory conditions by Johnson *et al.* (1964)—was in large-scale production by the end of 1966 by Rolls Royce Limited. This material was prepared in the form of 0.9 m-square sheets and used for the fabrication of aero engine components, each sheet having a moulding thickness of 0.127 mm with manufacturing tolerance of ±0.0127 mm. A fully continuous manufacturing process followed in 1968 (Whitney, 1973) and about this time a number of other manufacturers commenced carbon fibre production using PAN precursor fibres. Carbon fibre manufacture is now widespread.

2.4 PAN-BASED CARBON FIBRES

2.4.1 General Background

At the present time, most carbon fibres are manufactured from polyacrylonitrile precursors. The molecular structure of pure polyacrylonitrile is represented in Figure 2.11. The polymer chains are preferentially aligned with the axis of the fibre, but there is a lack of 3-dimensional order in the structure that can be attributed to the irregular distribution of the nitrile groups about the axis of the chain. The strong intermolecular forces due to the nitrile group do not allow the reorganization necessary for the development of crystallinity, so that the material is effectively amorphous (Goodhew, *et al.*, 1975). Because the material does not melt it has to be manufactured in fibre form from a solution of the polymer, i.e. by wet or dry spinning. In the dry spinning process the spinning solution is extruded through spinnerets into a hot chamber within which the solvent is evaporated. In wet spinning the solution of the polymer is squirted as a fine jet into a coagulating bath in which the polymer precipitates in the form of a fibre. Fibres are produced in tows which may be small, containing a few hundred filaments, up to tows containing about 10^4 filaments. The size of the individual fibres is less variable, the bulk of the fibres produced ranging from about 1 denier (~10 μm diameter) up to about 5 denier (~20 μm diameter). However, very much larger filaments are also manufactured.

The various proprietary types of PAN fibres can contain appreciable quantities of additional materials—typically methyl acrylate and vinyl acetate (Sanders, 1968). Fibres that are nominally similar can differ very significantly in terms of additives such as the by-products and traces of the initiator of the original polymerization, ionic groups that may have been introduced during spinning, co-monomers and residual solvent. These

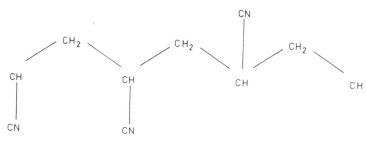

Figure 2.11. Representation of the structure of PAN.

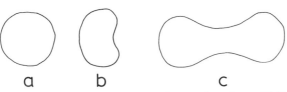

Figure 2.12. Illustrating typical cross-sections of various PAN textile fibres: (*a*) Courtelle; (*b*) Acrilan; (*c*) Orlon.*

substances can exert a considerable influence on the effects of thermal treatment and make it very difficult to develop a complete understanding of the processes that occur. In addition, the different spinning techniques used to manufacture the textile fibres influence the cross-sectional shapes of the fibres, which may be circular, kidney-shaped or of dog-bone form (Figure 2.12).

Polyacrylonitrile textile fibres exhibit two transition temperatures (Andrews and Kimmel, 1965; Rose, 1971), which differ between the various commercial products but are typically about 100°C and 160°C. Either of these could be described as a glass transition temperature. They can be observed as a discontinuity in the rate of change of specific heat with temperature. The preferential alignment of the molecular chains with the fibre axis can be improved by stretching the fibre at temperatures above its glass transition temperature. In many commercial textile PAN fibres a co-monomer carrying a bulky side-group is present, which reduces the glass transition temperature and also provides easier access for dyes (Goodhew *et al.*, 1975) and allows the fibres to be stretched in water at 100°C, thus increasing the degree of alignment of the molecular chains without initiating chemical degradation of the material. When the fibres are cooled under tension, the improved orientation of the molecular chains is preserved. Conversely, reheating in the absence of tension allows the molecular chains to revert to a more disordered state. Since the alignment of the graphitic "ribbons" in the carbon fibres is controlled by the initial alignment of the molecular chains in the precursor material, the production of high-elastic-modulus carbon fibres from PAN textile fibres requires the maintenance of tension at least during the early stages of pyrolysis. With increasing temperatures chemical changes take place in the textile material and extension or contraction to some equilibrium amount will occur, depending on the magnitude of the applied stress and the heating schedule that is followed. This is partly due to relaxation of the polymer chains and partly due to chemical changes taking place in the material (Rose, 1971; Müller *et al.*, 1971; Bromley, 1971).

---

* Courtelle is a trade name of Courtaulds plc. Acrilan is a trade name of Monsanto Fibres and Intermediates Co. Orlon is a trade name of E.I. DuPont de Nemours & Co. Inc.

A. E. Standage and R. Prescott observed that carbon fibres having high values of elastic modulus could be produced by heat treating various types of PAN textile fibres in a nominally inert atmosphere at very low initial rates of increase of temperature (<1°C per minute). The fibres were restrained from contracting during the initial stages of pyrolysis by being wound on a frame (Standage and Prescott, 1966). Elastic modulus values in excess of $350 \, GN \, m^{-2}$ were attained following further heat treatment at temperatures approaching 3000°C. These very slow initial heating rates were found to be necessary because of difficulties associated with partial melting and adhesion of fibres during the first stage of thermal treatment. Similar phenomena had been observed previously in the development of a process for the manufacture of a fireproof textile fibre from "Orlon"[*] polyacrylonitrile fibres by controlled pyrolysis (Vosburgh, 1960). With this material a sharp exotherm is developed at 221°C under inert atmospheric conditions, but controlled pyrolysis can be achieved by an oxidative heat treatment carried out in air in a forced-draught oven with fibre shrinkage being allowed during processing. The thermally stabilized fibre is blackened and becomes extremely resistant to subsequent oxidative degradation in air. Standage and Prescott (1965) observed that the elastic modulus of carbon fibres produced by the slow pyrolysis of PAN precursors could be modified by stretching the precursor fibre in water at 100°C, to improve the alignment of the molecular chains with the fibre axis. Subsequent contraction during pyrolysis was prevented by winding the fibres on a frame. It was found that the elastic modulus of the carbon fibres produced by a given set of processing conditions approached an upper limiting value with increasing extension of the polymer precursor. Young's modulus values of about $400 \, GN \, m^{-2}$ were observed. Slow initial rates of increase of temperature, e.g. $0.5°C \, min^{-1}$ were utilized during the first stage of thermal treatment and the beneficial effect of the use of an oxidizing atmosphere was noted.

Johnson *et al.* (1964) had earlier carried out an oxidative treatment of PAN textile fibres in air under tension. By varying the load applied to a bundle of fibres, elongation or contraction could be caused during the initial heat treatment. If the polymer fibres were allowed to contract during oxidation, the carbon fibres obtained after subsequent high-temperature heat treatment had relatively low values of elastic modulus. Conversely, if the fibres were prevented from contracting, or were extended during oxidation, much higher-modulus carbon fibres could be produced. By this means fibre elastic modulus values in excess of $400 \, GN \, m^{-2}$ were obtained (Figure 2.13). The bundles contained only a relatively small number of fibres and no problems of chemical or mechanical instability were encountered providing the initial heat treatment was of sufficient duration to permit complete permeation of oxygen throughout the individual fibres.

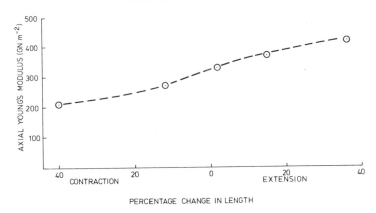

Figure 2.13. Length changes in PAN fibres during heat treatment in air at 220°C against elastic modulus after subsequent heat treatment at 2500°C. (Data from Johnson *et al.*, 1964.)

The tensile strengths of both oxidized and unoxidized polyacrylonitrile fibres fall as the temperature is increased and have minimum values at about 300°C, above which their strengths increase with increasing temperature (Watt, 1970). The loss of strength is less marked in the case of oxidized fibres.

The earlier work of Standage and Prescott was extended by Moreton (1971), who confirmed the direct relationship between the amount by which the textile precursor was stretched at 100°C and the elastic modulus of the carbon fibre produced after subsequent heat treatment. He also confirmed that there was a relationship between the temperature at which the textile precursor was stretched and the elastic modulus of the carbon fibre produced after subsequent pyrolysis. This was found to increase as the temperature increased at which the precursor fibre was stretched, reaching a limiting condition for temperatures of 150–160°C.

Earlier data on the properties of PAN fibres first oxidized (with shrinkage limited to about 5%) and subsequently heat treated to temperatures in excess of 2000°C were obtained by Shindo (1961, 1971). After high-temperature heat treatment carbon fibres having elastic moduli less than 200 GN m^{-2}, and with relatively low tensile strengths, were produced.

A preliminary oxidative heat treatment markedly improves the chemical and mechanical stability of the intermediate material during further pyrolysis to a wholly carbon fibre. This increased stability, coupled with the necessity to obtain a high degree of preferential alignment of the molecular chains of the textile precursor and to maintain this during processing, are the key requirements in a practical production process based on PAN precursors.

It was particularly convenient that high-tenacity experimental PAN fibres of small diameter (1.5 denier) in relatively small tow sizes (10,000 filaments) became available (from Courtaulds Limited) at the time that carbon fibre manufacture on an appreciable scale was started in the United Kingdom in 1966. The high degree of molecular alignment present in this type of fibre made it possible to produce high-elastic-modulus carbon fibres without the necessity of stretching the precursor. The small diameter of the fibres and the small tow size facilitated oxygen penetration and made it possible to avoid problems due to exothermic reactions during processing. The small tow size also facilitated the production of thin laminates of uniform thickness consisting of unidirectionally aligned fibres.

It is important to appreciate that, during normal oxidative processing, both "inert" and oxidative processes are proceeding at the same time, since the central regions of the fibres are being subjected to "inert" atmosphere heat treatment prior to the arrival of oxygen by diffusion from the surrounding atmosphere. Thus two markedly different polymer intermediates are present during the first stages of processing. Oxygen may be present in the co-monomers of some textile fibres, so that there exists the possibility, in some instances, of oxidative reactions taking place within the body of the fibre during heat treatment.

2.4.2 Oxidative and Inert Thermal Treatment of PAN Fibres

The properties of the carbon fibres produced by subsequent thermal treatment are very largely controlled by the processes taking place during the initial stabilizing step. This typically involves heating the PAN precursor fibres within the range 200–300°C under tension in air for 1 or 2 h. The processes occurring during this step can be deduced to some extent from an analysis of the gases produced, from thermal analysis, from changes in mechanical and optical properties and from x-ray data. These processes are complex and are not yet fully understood. (See e.g. Chen *et al.*, 1981; Jain and Abhiraman, 1983.) It is established that acrylic fibres are composed of sub units, termed fibrils, that are aligned parallel to the fibre axis during the drawing process. Warner *et al.* (1979b) have suggested that the fibrils are composed of rod-like structures that are aligned parallel to the axis of the fibril and that these are interspersed with less-ordered material.

As with other linear polymers, pyrolysis of polyacrilonitrile modifies the chemical structure and mechanical properties of the material. As the temperature is increased the colour of the fibres changes through yellow,

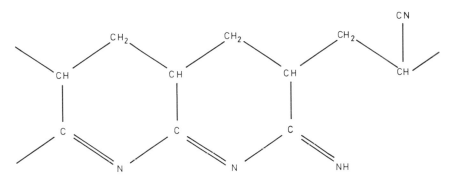

Figure 2.14. Illustrating the formation of a ladder polymer from PAN. (Redrawn from Watt, 1970.)

dark brown and black, and considerable contraction occurs if the fibres are not under restraint. The most generally accepted mechanism for the colour changes occurring during thermal treatment in an inert atmosphere is the building up of a ladder polymer by nitrile polymerization, as indicated in Figure 2.14. Neglecting initiation and termination points, this is a naphthyridine ring structure forming a "ladder" polymer (Grassie and Hay, 1962).

As the temperature is increased progressively in an inert atmosphere, a large reaction exotherm is observed at temperatures of about 300°C. Its magnitude and the temperature range over which it is developed depends on the particular type of PAN textile fibre. The exotherm is associated with the release of large quantities of ammonia (Watt *et al.*, 1974). An appreciable quantity of hydrogen cyanide is also released at this stage. In some types of polyacrylonitrile textile fibres the exotherm is very sharp, having the characteristic shape of an autocatalytic reaction. In other cases the exotherm starts at a much lower temperature and extends over a wide temperature range. The exotherm can be ascribed to the formation of the "ladder" polymer by the nitrile polymerization or cyclization theory, with NH_3 being released by the aromatization of the end ring of a conjugated sequence (Hay, 1968). The considerable increase in the thermal stability observed after pyrolysis would also be expected from the development of a "ladder" polymer because of the lower probability of both sides of the "ladder" rupturing thermally at the same time at the same "rung". The stability of the fibre is improved by preheating under inert conditions at temperatures below that at which the exothermic reaction occurs (Goodhew *et al.*, 1975). PAN fibres become insoluble in the solvents generally used to dissolve polyacrylonitrile very early during thermal treatment.

The intermediate fibre produced by oxidative heat treatment is very different from that produced by pyrolysis under inert atmospheric conditions. The practical importance of oxidation rests on the enhanced mechanical strength and thermal stability of the intermediate fibre produced. This considerably increases the carbon yield after subsequent processing, reduces the amount of subsequent fibre shrinkage and reduces the time required for carbonization. The enhanced intermediate fibre strength is also a major advantage during processing. Oxidized fibres are also intensely hygroscopic (Warner *et al.*, 1979c).

The principal chemical reactions that have been proposed as occurring during oxidation are outlined below. Firstly, it is found that during oxidation a polymerization reaction involving the nitrile groups occurs, with the formation of a ladder structure similar to that obtained during inert pyrolysis. It is also found that oxygen is introduced as OH and C=O groups that are assumed to be bonded to the carbon backbone of the PAN molecules (Clarke and Bailey, 1974). From infra-red spectroscopy it can be deduced that carbon double bonds are produced readily on oxidation, whereas in inert atmosphere high temperatures or long heating periods are required. At the same time, it has been shown from thermal analysis that chain splitting occurs during heating in an inert atmosphere but is absent in the presence of oxygen. Clarke and Bailey (1974) conclude that the main difference between oxidation and inert pyrolysis is the presence in the oxidized fibre of intermolecular hydrogen bonding, attachment of hydroxyl and carbonyl groups to the carbon backbone and considerable unsaturation and aromatization. These conditions are not apparent in the corresponding PAN fibre pyrolysed in an inert atmosphere. The oxygen-containing groups would be expected to promote cross-linking reactions and hence stabilize the precursor graphite layer planes in the material during the early stages of carbonization. In Figure 2.15 the expected chemical structure of oxidized polyacrylonitrile is illustrated.

The detailed changes occurring under thermal treatment depend on the particular material and processing conditions, but are broadly similar for various PAN textile fibres. A comparison has been made between two commercial PAN textile fibres by Watt *et al.* (1974). One was wet spun, with a diameter of about 13 μm (1.5 denier), and contained 6% methyl acrylate and about 1% of an acidic constituent. The other fibre was dry spun from nearly pure polyacrylonitrile and had a bean-shaped cross-section of 2.2 denier (16 μm equivalent diameter). Oxidized samples of both types of fibres were prepared, the oxygen uptake being 10.5%. A heating rate of $100°C\,h^{-1}$ was used during pyrolysis from 250°C to 1000°C for both

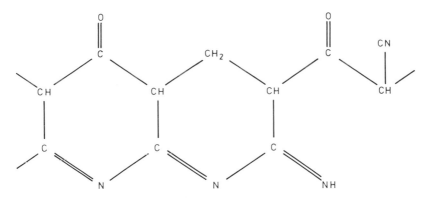

Figure 2.15. Oxidized PAN ladder polymer. (Redrawn from Watt, 1970.)

oxidized and unoxidized material, and the unoxidized fibres were given a preliminary heat treatment at 200°C *in vacuo* for 15 h.

The main gases evolved were NH_3 and HCN, both showing peaks at about 300°C and 700°C. The rate of evolution of gas was reduced in the case of both fibres after prior oxidation. At the higher temperatures nitrogen was released. The structural changes occurring during pyrolysis can be summarized as follows. About 40% of the unoxidized material and about 60% of the oxidized material is transformed into ladder polymer. The low figure for the unoxidized material, for the particular processing conditions used, can be accounted for by decomposition during the exothermic reaction taking place at about 275°C. This is suppressed in the oxidized samples. Breakdown of the non-laddered portions of the polymer follows and continues up to a temperature of about 500°C, but even with the unoxidized material some preferred orientation to the fibre axis is retained. End-to-end linkage of the laddered fragments begins between 400°C and 500°C and is visualized as incipient graphitic ribbon formation. In the case of the oxidized fibres subjected to the same rate of increase of temperature, ladder formation occurs at a lower temperature and to a greater degree, finally reaching about 60% conversion. There is also less extensive breakdown of the non-laddered fragments, which would be expected as a consequence of cross-linking reactions in this material.

During pyrolytic conversion both oxidative and inert atmosphere reactions are proceeding at the same time. Also, as the oxygen penetrates progressively into the inner layers of the fibres from the atmosphere outside, it reacts with precursor textile material that has been subjected to "inert" pyrolysis. The diffusion of oxygen may or may not be rapid compared with the rate at which pyrolitic reactions are taking place. If the diffusion rate is relatively slow, further diffusion of oxygen has to take place through material that has been pyrolysed under "inert" conditions and subsequently oxidized. All of these processes are influenced by the chemical composition of the material, the shape and area of the fibre cross-section and the fibre's temperature and thermal history.

The influence of these various factors have been studied by various investigators (Warner *et al.*, 1979a; Love *et al.*, 1975; Watt and Johnson, 1970) and it is observed that in the case of some types of acrylic fibres subjected to particular oxidative and thermal treatments a two-zone morphology is developed within the cross-section of the fibre. The outer zone is rich in oxygen and increases so as to encompass the whole cross-section of the fibre as the period of oxidation is increased. The outer layer becomes apparent after a finite time and is observed to have a particular thickness at this stage. Subsequently, the progress of this zone towards the centre of the fibre proceeds at a rate controlled by the diffusion of oxygen through the outer layer. Clearly, if the fibre radius is less than the outer oxygen-rich layer when it first becomes detectable, a two-zone morphology characteristic of a diffusion controlled process will not be observed. The thickness of this first-seen outer mantle provides a measure of the thickness of the material beyond which the reaction becomes diffusion-controlled.

This dimension will be influenced by a number of factors, including the fibre composition. For fibres of some particular compositions a two-zone morphology is not observed and oxygen is found to be distributed uniformly throughout the fibre cross-section as it reacts with the precursor fibre. However, if such fibres are subjected to a preliminary heat treatment *in vacuo* so that their chemical structure is modified, a two-zone diffusion controlled reaction can be observed during subsequent heat treatment in the presence of oxygen.

Warner *et al.* (1979a) describe these phenomena in terms of "prefatory reactions" that occur prior to or concurrently with polymerization of the nitrile groups and "sequent reactions" which occur subsequent to nitrile polymerization. When prefatory reactions occur significantly before the sequent reactions, the transport of oxygen to reactive sites has to take place through previously oxidized material and a two-zone cross-sectional morphology is developed. If the sequent reactions occur immediately after the prefatory reactions, the oxidation process is limited by the rate at which the prefatory reactions occur and a two-zone morphology is not developed.

The chemical composition of the fibre influences the initiation of nitrile polymerization; a weak acid co-monomer, which is present in some commercial acrylic fibres, is particularly effective in this respect. Hence, a fibre having a small diameter, with no weak acid co-monomer present and treated at relatively low temperatures, is not likely to develop a two-zone morphology. In contrast, large fibres containing a catalyst for prefatory reactions and treated at high temperatures are likely to develop a two-zone morphology during oxidation. Similarly, a preliminary heat treatment in an inert atmosphere serves to initiate nitrile polymerization so that again a two-zone morphology is more likely to occur during subsequent oxidation.

Warner *et al.* (1979a) suggest that an outer skin, having a thickness of the order of 1 μm, is also formed in the case of fibres in which a progressive increase in thickness of any outer oxygen-rich zone during processing is not observed. The presence of such a skin can be detected by etching the fibre and it acts as a fixed barrier to oxygen diffusion so that during oxidation the uptake of oxygen is proportional to the fibre surface area. The reason for the presence of this layer is unclear. It is also observed that tensile stresses developed in acrylic precursors during oxidative stabilization vary with time in different ways depending on whether the process is reaction-limited or proceeds by the progressive increase of an oxygen-rich outer layer.

The time required for atmospheric oxygen to diffuse to the central regions of the fibres will depend on a number of factors, as discussed above. Different regions of the fibre will experience different processing conditions during pyrolitic conversion of the fibres and this can result in differences in local material properties across the fibre cross-section. Optical interference phenomena observed by the reflection of polarized light from polished sections of carbonized fibres show effects that can be attributed to the development of internal stresses in differentially oxidized material; these techniques have been used in the study of circular sectioned fibres produced from a particular PAN precursor by Johnson (1979). At a temperature of 200°C the rate of chemical reaction of the PAN with oxygen was observed to be slow and the rate of oxygen diffusion sufficiently great to prevent the development of any significant oxygen gradient. At temperatures greater than 220°C the increase in reaction rate causes the oxygen penetration to be diffusion-controlled, so that a central zone containing little oxygen is developed which is surrounded by a much more highly oxidized sheath. From dark-field electron microscopy the layer planes of the graphitic crystallites produced after subsequent heat treatment are seen to be preferentially aligned parallel to the fibre axis but with no preferential radial or circumferential orientation of the layer planes within the core and the sheath zones. Hence, it is concluded that the differential stresses observed are generated not by gross structural anisotropy but by the greater contraction experienced by the less well oxidized core during thermal conversion to carbon.

Differential volume changes occurring in the initial polymeric material during conversion may generate voids, cracks or stresses in the material being processed. In addition the gaseous reaction products developed during conversion may cause the development of voids or cavities. Some precursor fibres develop a pronounced skin/core effect that is preserved in the carbon fibre that is eventually produced. In extreme conditions the much lower mechanical stability of the unoxidized core compared with the oxidized outer layers may lead to the production of hollow fibres during subsequent carbonization as a consequence of the volatilization of material within the core.

Further information on the changes induced in PAN during pyrolysis is available from X-ray data. When an X-ray beam is incident on a parallel bundle of fibres, orientated with its axis perpendicular to the beam, diffraction arcs are produced on a photographic plate orientated in a plane perpendicular to the beam direction. A central aperture is provided in the photographic plate through which an X-ray beam is passed so that only the diffracted portion is detected. The arcs formed can be related to particular distances of separation of the molecular chains in the polymer fibre, increased ordering being indicated by a reduction in the width of the arc. Only equatorial arcs are formed and this indicates that the material in the polymer fibre is well ordered in the two dimensions perpendicular to the fibre axis but has little preferred ordering in the direction along the fibre axis. The angular width at half-intensity of the equatorial peak of the arc is generally quoted as the X-ray orientation angle Z and is in the range $10–20°$ for most textile precursors. The width of the diffraction arcs in different PAN precursors can be attributed to the reduced ordering due to the presence in some cases of bulky side-groups on the polymer chains of the PAN textile fibres. The material appears closely packed between the side-groups with a chain-to-chain distance of about 0.6 nm, this accounting for the fairly high density of the polymeric material. In addition to the preferred separation distance of the individual polymer chains, there is some evidence of the existence of ordering on a much coarser scale, with assemblies of polymer chains having diameters of order 30 nm being present. During heat treatment a periodicity of about 0.35 nm becomes observable, apparently due to the stacking of planar cyclized polymer chain fragments as ladder structures are formed. As the temperature is increased, oxygen, nitrogen and hydrogen are driven off and graphite layer planes begin to form, the periodicity of the structure falling progressively to that of turbostatic graphite (~ 0.34 nm). During this stage of pyrolysis the structure of both oxidized and inertly treated PAN textile fibres, as deduced from X-ray data, are similar, but in the case of the oxidized fibre the interlayer spacing associated with the later development of the graphitic layer planes is more pronounced and becomes apparent at lower temperatures than is the case with the inertly treated material. These observations are in accordance with the occurrence of structural changes deduced from other data.

2.4.3 Carbonization of PAN-Based Fibres

Oxidized intermediate material is treated in an inert atmosphere, usually nitrogen, up to a temperature in the region of 1100–1200°C. This temperature is within the capability of a simple furnace. An additional stage, capable of reaching temperatures in the region of 1500°C is sometimes added to the carbonization furnace when fibres having rather higher values of elastic modulus are required. It is not normally necessary to apply mechanical restraint to the oxidized intermediate fibre during carbonization. Some longitudinal shrinkage occurs of the order of 10%. If tensile loads are applied to oxidized PAN fibres during subsequent carbonization, some elongation can be generated depending on the magnitude of the applied load. However, this elongation occurs at temperatures less than about 600°C and as the temperature is increased a small amount of shrinkage occurs that is independent of the magnitude of the tensile load. This phenomenon has been examined for a particular textile precursor under particular processing conditions by Bromley (1971).

As the temperature is increased above about 500°C similar changes occur in both oxidized and inert atmosphere pyrolysed PAN. HCN and N_2 are eliminated from the structure and the elastic modulus of the fibres increases with increasing temperature of heat treatment. Possible reactions taking place during the elimination of these gases are illustrated in Figure 2.16.

The nitrogen content of the fibres decreases with increasing temperature. The rate of release increases progressively up to 1000°C in the case of initially oxidized PAN fibres but reaches a peak at about 900°C for unoxidized material. The increase in elastic modulus of the fibres shows a strong correlation with the fall in nitrogen content, which is reduced to about 10% at 1000°C (Watt et al., 1974) falling to about 1% for heat treatment temperatures of 1300°C (Tyson, 1975).

In the case of both oxidized and unoxidized PAN fibres, the evolution of HCN takes place mainly in two temperature regimes, the first in the range 300–400°C, the second in the range 600–800°C. However, the magnitude of the HCN production peak is greater at the higher temperature for the oxidized fibre and greater at the lower temperature for the unoxidized fibre (Watt et al., 1974).

The loss of HCN and N_2 from the ladder polymer with increasing temperature of carbonization results in a lengthening and broadening of the pseudo-graphitic basal planes being developed in the fibre. X-ray data

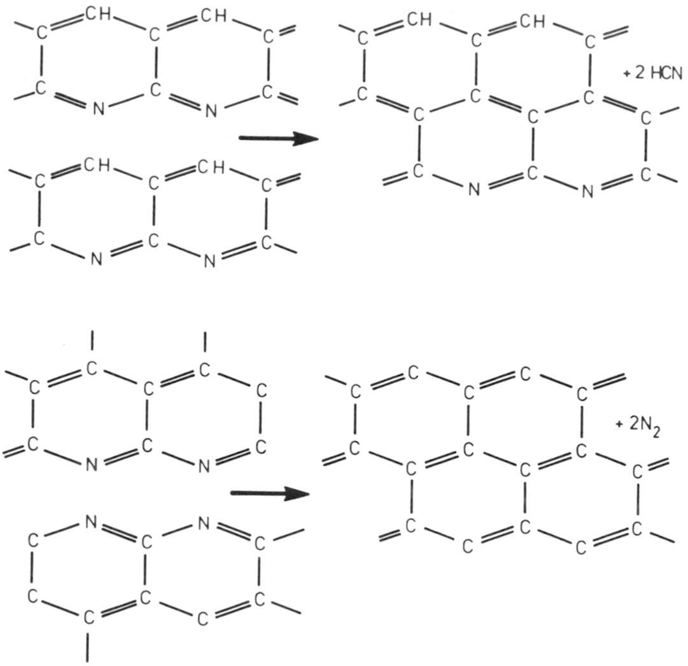

Figure 2.16. Structural changes occurring during the conversion of PAN fibres to carbon fibres. (Redrawn from Watt *et al.*, 1974.)

show that the length of these planes parallel with the fibre axis is growing more rapidly than the breadth normal to the axis during carbonization (Watt *et al.*, 1974). The lateral configuration of the layer planes in carbon fibres is discussed further in §2.5.2.

2.4.4 High-Temperature Heat Treatment of PAN-Based Carbon Fibres

Additional heat treatment at temperatures about 2500°C or even higher can be carried out to increase further the elastic modulus of PAN-based carbon fibres (Figure 2.17). This is often known as a graphitizing stage although, as explained above, the material produced is normally of a turbostatic form and not true graphite. If a tensile load is applied, fibre elongation is observed at temperatures in excess of 1700°C (J. W. Johnson *et al.*, 1969). The extension that can be generated depends on the processing temperature (Rose, 1971); extensions approaching 30% have been observed at temperatures of 2800°C, producing Young's modulus values of over 650 GN m^{-2}

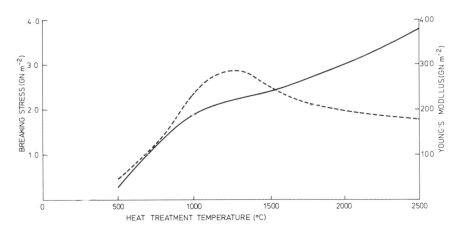

Figure 2.17. Typical Young's modulus (————) and strength (– – – –) values for early production PAN-based carbon fibres as a function of final heat treatment temperature. (Redrawn from Watt, 1970.)

(W. Johnson, 1970). This extension increases the elastic modulus of the fibre and the orientation of the c axis of the crystallites, and also the fibre tensile strength. These extensions are much less than those required of cellulose-based fibres to produce similar values of Young's modulus.

The Young's modulus of carbon fibres previously heat treated to 1500°C can be increased by allowing a small quantity of boron to diffuse into the fibres. This can be achieved by heat treating the fibres at elevated temperatures in an atmosphere containing boron. Very considerable increases in the Young's modulus are observed for a boron content in the region of 1% (Allen et al., 1970). The difference between the boron-containing fibres and the standard carbon fibres increases as the temperature is increased at which the fibres are exposed to boron. Elastic modulus values of $545\,GN\,m^{-2}$ were obtained after heating fibres containing 1% boron to a temperature of 3000°C, compared with values of $400\,GN\,m^{-2}$ obtained for the boron-free samples heated to the same temperature. It has been suggested that the effect of the boron is to enhance the recrystallization process in the fibres, especially the preferred orientation, and also increase the shear modulus of the material by solid solution hardening.

The effect of stretching is readily understood in terms of the graphitic ribbon model (Figure 2.10). The convoluted ribbons are preferentially orientated with the fibre axis and considerable entanglement is suggested. The spaces between the "ribbons" form irregular angular pores that reduce the fibre density from the single-crystal value of $2.265 \times 10^3\,kg\,m^{-3}$ to within the range $1.7–1.9 \times 10^3\,kg\,m^{-3}$ typical of carbon fibres. Although the broadening of high-angle x-ray lines is normally related to crystallite size, these data have to be interpreted differently in terms of the ribbon model. The dimension L_a represents the mean length of straight basal planes and not a discontinuous change in structure at boundaries spaced L_a apart. The average stacking height of the ribbons is represented by L_c. Both L_a and L_c increase with increasing temperature of heat treatment and this can be interpreted as an increase in the preferred orientation of the material. The L_c dimension for fibres heat treated to 2500°C is in the region of 5 nm, which corresponds to only a small number of interlayer spacings of turbostatic graphite. The values obtained for L_a are in the region of 10 nm (Goodhew *et al.*, 1975). The preferred orientation of the material is increased by hot stretching; this can be visualized as occurring by the collapse of the pore structure with the corresponding increase in degree of orientation and in the effective "crystallite" dimensions.

Quite different crystallization characteristics are developed if part of the textile precursor is allowed to fuse and volatilize during pyrolytic conversion. This condition is attained if the PAN fibre is oxidized for a short period at a fairly high temperature (~300°C), so that only the outer skin of the textile precursor is stabilized by oxidation (Johnson *et al.*, 1970). Following fairly rapid carbonization to 1000°C a central hole can be formed in the intermediate fibre, apparently due to partial melting and volatilization of the core material. This gives greater mobility to the material in this region in its structural development during subsequent heat treatment. Concentric lamellar graphitic structures begin to develop around the central hole following further heat treatment to temperatures in excess of 2000°C. The lamellar structure becomes more pronounced with increasing temperature of heat treatment and can extend as a hollow cylinder beyond the fracture face of a broken fibre.

2.4.5 Strength of PAN-Based Carbon Fibres

It was shown in §1.2 that the tensile failing strain of an ideal elastic solid would be expected to be about 10%. The failing strains of carbon fibres are very much less than this: the average failing strain of Type II high strength fibres currently being manufactured is about 2%. This represents a substantial increase over the failing strains obtained during the early development of carbon fibres. Fibre tensile strengths were observed to increase approximately linearly with increasing fibre elastic modulus, produced by increasing heat treatment temperature from 500°C up to temperatures of about 1200°C. Over this range of heat treatment temperatures the fibre failing strains were found to be fairly constant at about 1% (Morley, 1971). Average fibre strengths fall progressively as the heat treatment temperature is increased above about 1500°C to produce higher values of fibre elastic modulus. Thus, over this range of heat treatment conditions, the fibre failing strain falls as the fibre elastic modulus increases (Sharp et al., 1974). (see Figure 2.17). Fibres of high elastic modulus initially had failing strains in the region of 0.5%. Rather higher values of failing strain are now obtained with high-elastic-modulus fibres ($E \sim$ 400 GN m^{-2}) and recent studies indicate that failing strains in excess of 1% should be attainable with such fibres (Bennett et al., 1983).

Considerable differences exist between the elastic moduli and failing strains of carbon fibres produced by various manufacturers. The fibre strengths show a very wide dispersion about the mean value and increase as the sample gauge length decreases. These are the characteristics of a flaw-induced failure mechanism and flaws of various types originating in various ways are present in carbon fibres. Failure from surface flaws can be observed and internal flaws of various types can be identified in PAN-based carbon fibres (Sharp et al., 1974). These are generated by organic and inorganic inclusions in the precursor material. In many cases the flaws are diconic cavities having their axis of symmetry parallel to the fibre axis. The smaller the size of the flaw, the greater the frequency with which it appears. Very fine needle-shaped pores are also present. The diameters of the larger flaws are usually in the range 1–3 μm and small inclusions are sometimes present within them.

The structure of the fibre adjacent to the flaw is modified and can initiate fibre failure. Reynolds and Sharp (1974) suggested that misorientated crystallites can fail under the application of a tensile load by basal plane rupture. This is because the concentration of shear strain energy in a misorientated crystallite is not relieved by cracks parallel to the layer planes. A crack formed in a basal plane by this mechanism will propagate across the basal plane and, by transference of shear stress, through adjacent layer planes. In order that a crack formed in this way can cause fibre failure, either (a) the crystallite size must be large enough to contain a crack of critical size or (b) the crystallite in which failure is initiated must be sufficiently continuous with its neighbours for the crack to propagate. The first condition will not normally be fulfilled because the effective crystallite size in carbon fibres is much less than the critical flaw size (Bennett et al., 1983). The second condition can be satisfied by the presence of large crystallites sometimes found in the material that surrounds flaws in carbon fibres. The basal planes of these crystallites tend to follow the contours of the cavity (Sharp et al., 1974; Bennett et al., 1983). Not all holes contain a coating of graphitic sheets; where it occurs it seems to be caused by catalytic recrystallization due to the presence of a particle of metallic impurity.

When individual fibres are tested in bending, only a very small volume of material is subjected to high tensile and compressive strains. Under these conditions the effects of discrete flaws would be expected to be reduced and early observations (Jones and Johnson, 1971; Thorne, 1974) indicated that fibre surface strains of about 3% could be generated, which corresponded to maximum stresses of about $6.0\,GN\,m^{-2}$. More recent studies (DaSilva and Johnson, 1984) indicate that the fibre strengths in bending are similar to the values observed under tensile loading. These authors also suggest that the Reynolds–Sharp failure mechanism operates in both tension and compression.

An investigation of the effect of foreign inclusions within the precursor material and the surrounding atmosphere during textile fibre manufacture has been carried out by Reynolds and Moreton (1980). They compared the characteristics of carbon fibres prepared from precursors manufactured under particularly clean conditions with fibres prepared from similar precursors manufactured under normal laboratory conditions. In the former case a series of filters reduced the particle count in the air surrounding the fibre spinning apparatus to a level almost four orders of magnitude lower than that measured for the control experiment conditions. In addition, the fibres were spun from either a normal polymer solution or one which had been filtered so as to remove particles larger than 1.5 µm from the solution.

These manufacturing conditions result in considerable differences in the tensile strengths of carbon fibres produced after oxidation, stabilization and subsequent carbonization. The differences in strength become more apparent the higher the temperature of subsequent heat treatment. This is illustrated in Figure 2.18, in which the data refer to the tensile strength of samples of fibre 5 cm long. It is clear that the reduction in tensile strength normally observed for PAN-based fibres heat treated to temperatures in excess of about 1500°C does not occur, for a 5-cm gauge length specimen, if appropriate precautions are taken to reduce the prevalence of flaws in the precursor material. Further confirmation that the reduction in strength of carbon fibres heat treated to high temperatures is due to impurity particles was obtained by Reynolds and Moreton by deliberately contaminating the clean textile precursor with carbon black, silica particles and particles of ferric oxide. Carbon fibres prepared in this way show a similar reduction in strength with increasing temperature of heat treatment to that of fibres prepared from standard textile precursors. It would seem that average strengths of about $3\,GN\,m^{-2}$ can be obtained over a 5 cm gauge length for high-elastic-modulus carbon fibres heat treated to 2500°C—providing careful precautions are taken to exclude foreign particles from the precursor material. As discussed above, the mechanism of flaw development from impurity particles is likely to be the growth of localized randomly orientated 3-dimensional graphite. This seems to occur by the dissolution of carbon

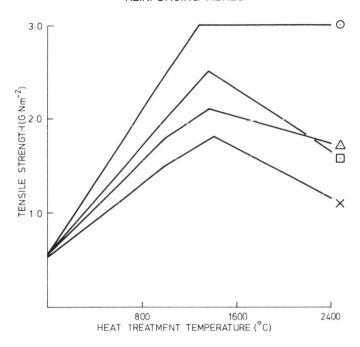

Figure 2.18. Effect of heat treatment temperature on the average fibre tensile strength of various samples. Filtered–clean, ○; filtered–normal, △; unfiltered–clean, □; Unfiltered–normal, +. (Redrawn from Reynolds and Moreton, 1980.)

into the impurity particle followed by the precipitation of graphite at higher temperatures, at which the metal carbide is unstable.

If the mode of failure of carbon fibres is governed by the presence of discrete flaws of varying severity, the observed average strength of the fibres should increase as the physical size of the samples tested is reduced. This effect was observed by Reynolds and Moreton (1980). Although fibre strengths measured for a gauge length of 5 cm show very considerable differences according to the conditions of manufacture, this is not the case when shorter gauge length fibres are tested. This point is of major importance in considering the effective strength of fibres when they are encapsulated in a matrix. Discontinuous fibres can be used effectively to reinforce a matrix, the parameters of interest being the aspect ratio of the fibres and the efficiency of stress transfer between the fibres and the matrix (see §§3.11 and 5.1). These factors have to be taken into consideration in assessing the influence of flaws on fibre strengths.

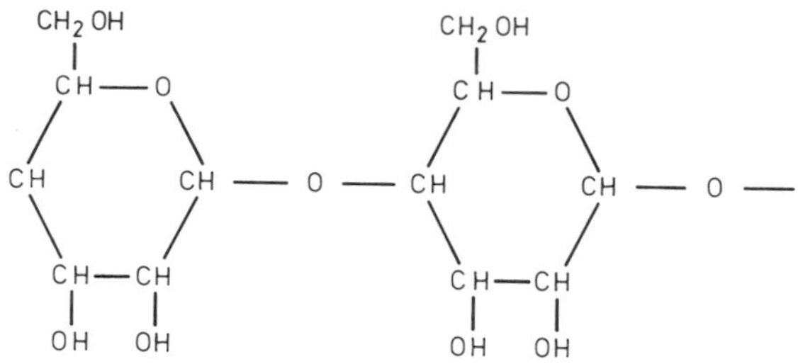

$$CH_2OH \qquad\qquad CH_2OH$$

Figure 2.19. General molecular form of cellulose.

2.5 CARBON FIBRES FROM CELLULOSIC PRECURSORS

Cellulose is a naturally occurring polymer, often existing in the form of fibres such as those of cotton. The chemical structure of the material is as indicated in Figure 2.19. The chain length is dependent on the source of the cellulose. For example, it is of the order of 10^4 repeat units for cotton. Rayon fibre manufacture utilizes wood pulp, the least expensive form of cellulose. Continuous filaments are produced by a wet spinning process, the polymer chains for this material consisting of several hundred repeat units. Each unit of the polymer chain contains hydrogen and oxygen corresponding to five water molecules and it follows, therefore, that the minimum weight loss during the conversion of pure cellulose to carbon is almost 56%. Naturally occurring cellulose has a structure known as cellulose I, and synthetic cellulosic materials have a similar crystal structure known as cellulose II.

2.5.1 Pyrolysis of Cellulose-Based Carbon Fibres

The conversion of rayon into high-elastic-modulus carbon fibres has been described in detail by Bacon (1973). It comprises three basic steps—preliminary heat treatment, carbonization and graphitization. A typical continuous process for carbonizing rayon has been described by Ross (1968). The fibre has to be stretched by an appreciable amount during the graphitization stage in order that high values of elastic modulus may be obtained. This is necessary because, in the preliminary heat treatment process, the preferred orientation of the cellulosic polymer structure within the precursor is destroyed during the formation of the primary char. Carbonization is usually carried out in an inert atmosphere to temperatures in the range 1000–1500°C. The preferred orientation of the graphitic structure eventually produced is then developed at high temperatures (in the region of 2500–3000°C) by tensile plastic deformation of the fibre.

During preliminary heat treatment the cellulose fibres are heated to temperatures of about 200–400°C and the treatment can be considered to occur in four stages.

Stage 1. Physical desorption of water (25–150°C).

Stage 2. Dehydration from the cellulose unit (150–240°C).

Stage 3. Thermal cleavage of the glycosidic linkage, scission of other carbon oxygen bonds and some carbon–carbon bonds via a free-radical reaction (240–400°C).

Stage 4. Aromatization (400°C and above).

These processes have been discussed in detail by Tang and Bacon (1964). The precursor fibres contain typically about 10% of physically absorbed water and this is removed during Stage 1. The avoidance of excessive tar formation is the most important single consideration in the design of the initial heating process which, besides enhancing weight loss, can cause interfilament cementation and surface deposits on fibres in yarns, leading to low fibre strengths.

When pyrolysed in a vacuum or inert atmosphere cellulosic materials yield volatile products consisting mainly of aqueous distillates, tars and various fixed gases. Under these conditions the weight loss from the fibre can be as high as 90%; hence very slow heating rates are necessary, e.g. ~10°C h^{-1}. Processing rates and overall weight yields are improved if the initial heat treatment is carried out in reactive atmospheres such as oxygen, air, chlorine and hydrochloric acid vapour. These substances all seem to promote dehydration of the cellulose structure and inhibit the formation of tars. Compared with pyrolysis in an inert atmosphere, dehydration reactions are initiated at a lower temperature and take place over a wider temperature range. The effectiveness of reactive atmospheres in increasing processing rates and weight yields is limited by the diffusion of the gas to the fibre interior which is, of course, temperature-dependent. Advantages can be gained by impregnating the fibres prior to heat treatment with aqueous solutions of materials that were initially developed to limit the flammability of rayon textiles when heated in air. These operate by reducing the availability of the hydroxyl groups necessary for the formation of laevo-glucosan (Goodhew *et al.*, 1975; Tang and Bacon, 1964). By these means heat treatment times of the order of 1 h become practicable in continuous manufacturing processes. Carbon yields after graphitization approach 30%; this corresponds approximately to a yield of four carbon atoms per cellulose ring unit.

Cellulose retains its crystalline structure up to about 245°C and at this temperature a weight loss of 15% or 20% has occurred. Above this temperature the cellulosic crystal structure starts to break down, appearing from x-ray diffraction data to be completely amorphous at temperatures over about 300°C. The tensile strength of the pyrolysed fibre falls during initial heat treatment, reaching its lowest value near the point at which the destruction of the crystalline cellulose is complete. As the temperature is increased further, however, the tensile strength of the pyrolysed fibre starts to increase. There is a slight increase in the Young's modulus of the fibre during the early stages of heat treatment, but thereafter it falls with increasing temperature of heat treatment and continues to fall after the fibre strength has started to increase. The elastic modulus starts to increase again at higher temperature as the graphitic layer structure begins to form.

Lateral and longitudinal shrinkage of the fibres occurs during the initial pyrolysis as material is being lost from the precursor fibre and it is impracticable to apply tension to the fibres during this stage of the process. The longitudinal contraction has been observed to diminish as the molecular orientation of the precursor fibre is increased by mechanical stretching during fibre formation, falling asymptotically to a value of about 25% with increasing degrees of initial stretch in the precursor. The magnitude of the shrinkage agrees well with a mechanism of longitudinal polymerization of cellulose ring unit residues, in which four carbon atoms remain, into graphite layers (Bacon and Tang, 1964). The carbonization stage removes almost all of the residual volatile material in the fibre and increases its tensile strength. Carbonization up to temperatures of about 1500°C can then be achieved within a period of less that 1 min and the fibre may be stretched during this stage to increase the degree of preferred orientation of the graphitic structure. This produces a modest increase in the Young's modulus of the carbonized fibres and the effect of stretching under these conditions is much more apparent if the fibre is subsequently heated to "graphitizing" temperatures without being subjected to further stress. This reorientation process is most effective if it is carried out during the early stages of carbonization when the fibre is developing a graphitic structure but is still quite plastic. By this means, values of elastic modulus of $200\,\mathrm{GN\,m^{-2}}$ can be developed. However, much higher values of Young's modulus can be obtained if the deformation is carried out at high temperatures. This can be done at temperatures in excess of 1600°C but is more conveniently carried out at temperatures of about 2800°C, at which the fibres become highly plastic. The Young's modulus of the fibres is observed to increase with the amount of fibre elongation during graphitization and also with the graphitization temperature.

This process results in a dramatic increase in the fibre elastic modulus and tensile strength. The elastic modulus developed in the fibre tested at room temperature increases progressively with increasing elongation at high temperature. An elastic modulus of about 620 GN m^{-2} can be developed by a fibre elongation of 400% at 2750°C, which causes a decrease in the fibre diameter by a factor of 2. The elongation required to develop similar values of Young's modulus in PAN-based carbon fibres is much less than this because of the initially higher degree of crystallographic orientation in this material.

2.5.2 Physical Characteristics of Cellulose-Based Carbon Fibres

As with PAN-based carbon fibres the Young's modulus of the fibres produced from rayon can be correlated with the degree of preferred orientation of the graphitic layer planes with the fibre axis. There is a close similarity between the carbon fibres produced from either precursor when the orientation half-width obtained from x-ray data is plotted against the Young's modulus of the fibres, particularly if a correction is made for the slightly lower density of the rayon-based fibres. The agreement occurs despite the fundamental difference between the manufacturing conditions used, owing to the maintenance of the preferred molecular orientation in the PAN precursor during processing and its destruction during the early stages of pyrolysis in the case of rayon. Thus, in order to achieve high values of elastic modulus it is necessary to hot stretch rayon-based carbon fibres.

The structure of the high-elastic-modulus fibres produced from either precursor is essentially similar, both types containing highly orientated, slightly convoluted ribbons of graphitic layer planes (see Figure 2.10). This implies a distribution of highly elongated pores within the fibres, having widths of the order of 1 nm and with lengths greater by one or even two orders of magnitude. The ribbons appear to be corrugated sheets that bend and curve through the transverse section of the fibre (Bacon 1973; Bennett and Johnson, 1978). This is illustrated schematically in Figure 2.20. The transverse widths of the layer planes are therefore much greater than is implied in the schematic longitudinal section shown in Figure 2.10. Hence the basic unit of the carbon fibre structure (analogous to long-chain molecules in polymers) is the graphitic layer. These layers are considered to be relatively straight and closely aligned with the longitudinal axis of the fibre but highly convoluted in the transverse direction. The sheets are arranged as turbostratic layers forming a continuous branching network. These connections occur both longitudinally and laterally, the frequency of branching and curvature of the graphitic layers being very much greater in a transverse section than in a longitudinal section. This structure can be considered as an array of microfibrils (Bacon, 1973) containing elongated micropores that are preferentially aligned with the fibre axis and possess irregular cross-sectional shapes. Preferential alignment of the microfibrils

Figure 2.20. Schematic representation of graphitic ribbons viewed in cross-section (After Bacon, 1973.)

with the fibre axis by stretching at high temperature increases the axial Young's modulus and reduces the degree of porosity of the fibre, thus increasing its density. This appears to occur not by a reduction in the average pore diameter but by a reduction of the total number of pores present by the collapse of some of them.

A number of investigators have proposed that a larger structural unit termed the fibril is present in carbon fibres (see Goodhew *et al.*, 1975; Bacon, 1973). Their transverse dimensions are estimated to be generally within the range 10–100 nm. The main evidence for the existence of fibrils comes from the irregular fracture surfaces of the fibres. The lateral boundary between two fibrils is thought to consist of graphitic layer planes across which relatively few connecting graphitic layers pass. It seems probable that the fibrils also form a continuous branching network. Studies of high-elastic-modulus carbon fibres produced from rayon precursors indicate that their structural details are similar to those of corresponding high-elastic-modulus fibres obtained from PAN precursors by normal manufacturing processes. As discussed in §2.4.4, pronounced concentric lamellar structures can be developed in PAN-based carbon fibres under some processing conditions.

The application of a tensile stress to a fibre causes a slightly misaligned array of graphite layer planes within a "ribbon" to be placed into longitudinal extension and also into shear deformation. The misaligned ribbon is tilted elastically towards the fibre axis as a tensile load is applied. Ruland (1969) has developed an analytical treatment based on this simple model by which the distribution function for the misalignment of the graphite layers can be related to the fibre Young's modulus for other precursor fibres in addition to rayon and for a wide range of fibre moduli and processing conditions.

The influence of flaws in the precursor material and the effect of surface irregularities have not been reported as extensively as in the case of PAN-based fibres. However, similar mechanisms would be expected to operate. The tensile strengths of stress-graphitized cellulose-based carbon fibres are almost directly proportional to their Young's modulus at relatively low values of strength and stiffness but depart progressively from this relationship as the Young's modulus is increased (Figure 2.21). Thus the failing strain is about 1% at an elastic modulus value of 140 GN m^{-2}, falling to about 0.5% when the Young's modulus has reached 690 GN m^{-2}. It is interesting to note that both high-elastic-modulus-stress-graphitized rayon-based carbon fibres and high-elastic-modulus PAN-based carbon fibres possess similar strengths, with average failing strains in the region of 0.5% (Figure 2.21).

Rayon-based carbon fibres are reported by Bacon (1973) to show a non-linear elastic response in bending when the peak strain exceeds about

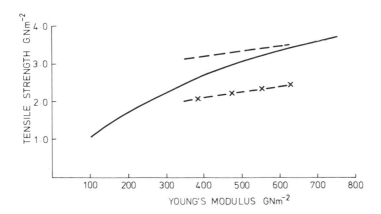

Figure 2.21. Relationship between tensile strength and Young's modulus for stress-graphitized rayon-based carbon fibres of 2-cm gauge length (Bacon, 1973), compared with stress-graphitized PAN-based fibres of 1 cm (———) and 5-cm (– × – – ×) gauge lengths (Johnson, 1970).

0.5% Beyond this value, buckling is observed within that portion of the fibre that is subjected to compressive stresses, analogous behaviour to that of a steel wire rope. Similar observations were made by Jones and Johnson (1971) of the behaviour of PAN-based fibres loaded in flexure. This phenomenon has not been observed to occur in more recent studies of the flexural behaviour of various PAN-based carbon fibres (DaSilva and Johnson, 1984). DaSilva and Johnson found the calculated failing strengths in bending to be very similar to the corresponding tensile strength values. Clearly, the behaviour of carbon fibres in compression will be governed by the microstructure of the fibres.

As is the case with PAN-based carbon fibres, the thermal expansion coefficient of cellulose-based fibres is slightly negative over the temperature range from room temperature to a few hundred degrees Celsius.

2.6 CARBON FIBRES FROM PITCH PRECURSORS

2.6.1 Hot-Stretched Pitch-Based Carbon Fibres

Pitch, produced as a by-product of the petroleum and coal coking industries, forms an inexpensive precursor for the manufacture of carbon fibres. Initially, high-elastic-modulus carbon fibres were produced from pitch precursor fibres by hot stretching techniques similar to those used in the manufacture of high-elastic-modulus fibres from cellulose precursors (Baker *et al.*, 1970; Hawthorne, 1976). After first refining the pitch by distillation, the precursor fibres are manufactured simply by extruding the molten pitch through nozzles of the appropriate size in a manner similar to that for glass fibres, the fibre diameter being controlled primarily by the melt temperature and the drawing speed. High stretch ratios can be used. It is necessary to convert the pitch fibres prepared in this way into a non-fusible state so that carbonization can be carried out at high temperatures without melting the fibres. The viscosity of the pitch, and hence the temperature and speed of spinning, is controlled by the molecular weight of the pitch, which can be adjusted by solvent extraction, by promoting condensation reactions between the molecules, or by adding a plasticizer to the melt. The low-strength fibres prepared in this way can be rendered infusible by controlled oxidation treatments that promote cross-linking reactions. The fibres can then be made thermally stable by further heat treatment at temperatures of about 1000°C.

The carbon fibres produced at this stage have modest average tensile strengths ($\sim$1.75 GN m^{-2}) and low elastic modulus values (20–50 GN m^{-2}). The size of the crystallites after heat treatment to 2500°C is in the range 3–5 nm with no measurable preferred orientation. These fibres can be extended under a tensile load at temperatures in the range 2000–2800°C. Elongations in excess of 100% can be developed, resulting in a considerable increase in the preferred orientation of the crystallites and a corresponding increase in the axial Young's modulus of the fibres. After hot stretching the room temperature strengths and Young's modulus values approach $3 \, GN \, m^{-2}$ and $440 \, GN \, m^{-2}$ respectively.

2.6.2 Carbon Fibres from Mesophase Pitch

The technologically inconvenient hot stretching stage can be avoided by first converting the precursor pitch into a liquid-crystal structure or mesophase. During extrusion to form fibres, the liquid-crystal structure becomes aligned with the fibre axis, so that a preferentially orientated graphitic structure can be developed in the fibre following further heat treatment. This process is based on investigations carried out by Brooks and Taylor (1965, 1968) on the mechanics of the formation of graphitizing carbons from the liquid phase. Substances that form graphitizing carbons in this way include high-temperature coal-tar pitch, petroleum bitumen and polymers such as polyvinylchloride.

During heating these substances melt to form isotropic pitch-like materials of various viscosities, typically having a treacle-like consistency. During heat treatment, hydrogen is lost from the low-molecular-weight aromatic molecules of the pitch and they join together forming larger 2-dimensional molecules. With increasing temperature, small spherical domains separate out from the pitch matrix. At the earliest detectable stage of their growth the material forming the spheres can be shown to be an anisotropic liquid, of slightly higher density than the surrounding isotropic pitch, possessing a lamellar structure, i.e. that of a liquid crystal. Brooks and Taylor obtained a value of 0.347 nm for the interlamellar spacing in the droplets and this is very similar to the interlamellar spacing of turbostratic graphite (~0.34 nm). With increasing temperature, and to some extent with increasing time at a fixed temperature, the spheres increase in size. Eventually the spheres become large enough to interfere with each other and the remaining isotropic material is absorbed by the liquid-crystal mesophase. The mesophase (so termed because it is an intermediate phase occurring between the precursor material and the solid graphitic material eventually produced) behaves as a liquid of high viscosity. As the temperature is increased further, the viscosity of the mesophase increases more or less rapidly depending on the particular precursor material.

During the early stages of their growth the mesophase droplets appear to form perfect spheres as a consequence of surface tension forces. They can be nucleated by the presence of fine-particle inclusions but these are not incorporated into the spheres as they grow. For a given thermal history, a pitch containing inclusions contains a large number of small spheres of mesophase material. The identical material without inclusions contains a smaller number of larger spheres. The total proportion of mesophase is observed to be approximately the same in both cases.

The chemical constitution of the pitch-like materials from which the spheres form is very complex. These are mainly polynuclear aromatic hydrocarbons (Brooks and Taylor, 1968). Brooks and Taylor suggest that during thermal treatment the more active hydrocarbons in the pitch phase are converted by dehydrogenation and condensation to larger, more complex species that are planar. These compounds then separate out from the pitch in partly ordered lamellar molecular assemblies to form the mesophase. There appears to be a lower temperature limit below which spheres of mesophase do not form. For coke-oven pitch this is about 400°C. At higher temperatures spheres form and increase in size with time; the higher the temperature, the shorter the time required for complete conversion to mesophase. When the spheres grow and coalesce the structure formed is described as mosaic. The planar molecules within the spheres are distorted near the poles when initially formed. Recent studies (Higuchi *et al.*, 1984) indicate that in these regions they cut the surface of the sphere at angles of 60–70° (Figure 2.22). When spheres of mesophase cohere, a rearrangement of the lamellae occurs at the interface so that continuous links are formed between the two spheres. Within the central regions of a sphere, a layer type of alignment is initially developed but the coalescence of two spheres can produce a complex internal arrangement, as indicated (in Figure 2.22). As coalescence of the spheres becomes general and the remaining isotropic material is absorbed, a 3-dimensional structure is formed in which local regions (wherein the orientation is fairly uniform) are

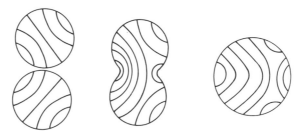

Figure 2.22. Schematic representation of the type of structure formed after the coalescence of two or more mesophase droplets. (After Brooks and Taylor, 1965.)

separated by zones in which the lamellae curve sharply in order to conform to the orientation of the next fairly uniform region. If they are held for long periods in the liquid condition the size of the ordered regions increases but, because of the very viscous nature of the mesophase, this process proceeds at a very slow rate. This progressive reorganization will not occur if isotropic pitch is still present, because of the strong control exerted by the mesophase–isotropic pitch interface on the ordering of the lamellae within the mesophase material.

During further heat treatment to higher temperatures the mosaic texture and lamellar orientation remain and, at temperatures above 2500°C, x-ray diffraction patterns indicate the development of 3-dimensional graphite structure. This is accompanied by anisotropic shrinkage, the effect of which can best be seen from the behaviour of a mesophase spherulite after being extracted from the surrounding isotropic pitch and heated to a high temperature. The sphere assumes an oblate shape owing to preferential shrinkage perpendicular to the layer planes within it. This shrinkage is not due primarily to a reduction in the interlaminar spacing to that of graphite, which is quite small, but seems to be associated with a large increase in density of the material. The density of the mesophase is reported by Brooks and Taylor (1968) to be only $1.4 \times 10^3 \, kg \, m^{-3}$, compared with $-2.3 \times 10^3 \, kg \, m^{-3}$ for perfect graphite. The layer planes contain large voids and it seems very probable that the voids are filled by the migration at high temperatures of the interstitial carbon atoms to form perfect 3-dimensional hexagonal arrays. Brooks and Taylor (1968) also noted that when the pitch was partially oxidized its conversion to a mesophase was strongly inhibited.

The mesophase can be formed into fibres by melt spinning at an appropriate temperature and this process produces a preferential alignment of the planar molecules with the fibre axis (Bacon, 1979). This occurs as a consequence of the longitudinal velocity gradients developed by the convergent flow of the liquid at the approach to the orifice. Radial velocity gradients are also experienced within the nozzle. The 2-dimensional molecules may be aligned in a radial fashion about the fibre axis, or be aligned circumferentially to produce an "onion-skin" type of structure; or the alignment may be random in the plane of the fibre. In all cases, however, the molecular layers are aligned with the fibre axis.

This preferential alignment within the fibre is stabilized by heat treatment in an oxygen-containing atmosphere at temperatures between 250°C and 400°C; this promotes further cyclization and cross-linking. During subsequent thermal treatment to temperatures in the range 2000–3000°C, perfection of the parallel-stacked crystallites increases, causing an increase in the elastic modulus of the fibre. Young's modulus values of about $800 \, GN \, m^{-2}$ can be obtained (Volk, 1977). A carbon yield in excess of

80% of that present in the fibre as initially spun is retained in the carbon fibre after final heat treatment.

It is observed that, when the layer planes of the mesophase pitch are arranged radially in the fibre, a radial crack can form during heat treatment. This can open up into a wedge during subsequent contraction of the fibre from a density of 1.3×10^3 kg m^{-3} when formed from mesophase material up to 2.0×10^3 kg m^{-3} after graphitization. These characteristics are deleterious in a reinforcing fibre but are not apparent in fibres in which the graphitic layer planes are arranged randomly in the fibre cross-section or in fibres in which the graphitic layers are arranged circumferentially. Here any voids produced by contractions occurring during heat treatment must be small and localized. The denisty of carbon fibres produced from mesophase pitch is greater than that of fibres produced from other precursors and the size of the crystallites within the fibres is also greater. As with fibres produced from other precursors, mesophase pitch-based fibres are observed to fail from surface flaws and from internal voids and foreign-particle inclusions. When special precautions are taken to minimize contamination, fibre tensile strengths in excess of 3 GN m^{-2} can be obtained with short gauge length (3.2 mm) specimens (Jones et al., 1980).

2.7 SURFACE TREATMENT OF CARBON FIBRES

During early investigations of the properties of composites containing a unidirectional array of carbon fibres it was observed that the values of longitudinal shear strengths were very low (~ 20 MN m^{-2}), although the axial tensile strengths of the composites were high (~ 1.5 GN m^{-2}). The low values of shear strength were due to poor bonding between the fibres and the matrix. It was observed that there was a pronounced inverse relationship between the elastic modulus of the fibres and the shear strength of the polymeric matrix composites produced from them. A comparable relationship exists with the size of the graphitic crystallites in the fibres, since the fibre elastic modulus is related to these values (§§2.4.4 and 2.5.2). An extensive account of the various factors that have been found to affect the nature of the interfacial bond in carbon-fibre-reinforced composites has been given by Plueddmann (1974) and this topic has been discussed more recently by Ehrburger and Donnet (1980).

The highly anisotropic nature of the graphitic crystal structure leads to two very different types of local surface characteristics, since the crystallites at the fibre surface may be aligned in various directions. The bonding energy is high in the basal plane of the crystal, so that the surface reactivity is low. Conversely the reactivity is high at the edges of the basal planes (Clarke et al., 1974). Improved interfacial bonding is obtained by oxidation simply by heating in air or by using liquid-phase techniques. The surface area of the fibres is increased by these processes and composite shear strength values are increased up to a limiting value.

Air oxidation is characterized by the growth of etch pits on the fibre surface. These are observed to be formed in lines that subsequently form channels as the etch pits increase in size and degree of coalescence. During this process the fibres show a rapid increase in surface area by about a factor of 4 for a fibre weight loss of about 1%. Fibre–matrix interfacial shear strengths can be increased substantially simply by immersing the fibres in a sodium hypochlorite solution at room temperature. This treatment produces surface pitting, with the subsequent development of channels similar to those generated by air oxidation. The interfacial shear strength is also increased, the value again rising to a limiting value with increasing duration of treatment. Care has to be taken to clean the fibre thoroughly after treatment and before composite fabrication.

2.8. SILICON CARBIDE FIBRES

Silicon carbide (SiC) fibres are of technical interest because of their good mechanical properties, resistance to oxidation and high temperature stability. The density of SiC ($3.21 \times 10^3 \, kg \, m^{-3}$) is rather higher than that of other advanced reinforcing fibres such as boron and carbon. Three types of SiC fibres are considered here: those produced by (a) chemical vapour deposition onto a tungsten wire or carbon filament; (b) the thermal conversion of a spun precursor polymer fibre; and (c) the manufacture of SiC whisker crystals from rice hulls and also by the VLS (vapour–liquid–solid) process. These processes and the properties of the fibres produced by them have been reviewed by Andersson and Warren (1984) and Warren and Andersson (1984). Whisker crystals are very small high-purity single-crystal fibres. They can be produced with smooth surfaces and, as a consequence, have very high tensile strengths. Provided their length-to-diameter ratios are high, their reinforcing efficiency approaches that of continuous fibres.

2.8.1 Fibres Produced by Vapour Phase Deposition

The process is very similar to that used in the manufacture of boron fibres. Initially a tungsten wire heated electrically was used as a substrate, but more recently a carbon filament has been used for this purpose. The silicon carbide is usually deposited from vapour mixtures of alkylsilanes and H_2 to produce a filament having a diameter between 100 µm and 150 µm. The deposition rate is controlled by the deposition temperature and factors such as the pressure and composition of the vapour mixture. At lower deposition rates the fibre structure is amorphous and the tensile strengths and Young's modulus values are similar to those of boron fibres. The optimum deposition temperature for commercial fibres appears to be about 1300°C, at which the crystallite size is about 30 μm and a 100 μm filament can be produced at a rate of about 5 m min^{-1}. High deposition rates can be achieved at tempertures above 1300°C, but a coarse crystalline fibre is produced and the fibre is weak. High fibre rupture strengths can be maintained up to temperatures of about 1100°C (Andersson and Warren, 1984).

The surface layers of silicon carbide filaments are readily oxidized by heat treatment and the oxidized hydrated surface possesses properties similar to those of glass fibres. Silane coupling agents, as are used with glass fibres, increase the interfacial bond with an epoxy resin matrix.

2.8.2   Fibres Produced from Melt Spun Precursor Fibres

Silicon carbide fibres produced by the thermal conversion of a polycarbosilane precursor fibre have been described in various publications—see e.g. Yajima et al. (1978), Hasegawa et al. (1980) and Hasegawa and Okamura (1983). The fibres are primarily composed of silicon carbide in an amorphous or crystalline form and contain appreciable quantities of SiO_2 and carbon. As a consequence, the fibre elastic modulus is relatively low, the generally accepted value being about 200 GN m^{-2} compared with values of about 400 GN m^{-2} for bulk full-density polycrystalline SiC. Some investigators have reported appreciably lower values of fibre elastic modulus (e.g. Simon and Bunsell, 1984a).

Various processes for the production of polycarbosilane have been investigated and this material can be used to manufacture the precursor fibre by melt spinning at 350°C in nitrogen. The precursor fibre is stabilized by oxidation in air at temperatures in the region of 200°C. Further heat treatment to higher temperatures is carried out in nitrogen or in vacuo. During heat treatment over the range 550–850°C decomposition occurs of side-chains carrying hydrogen and methyl groups. The density of the fibre increases rapidly at 600°C from an initial value of about 1.3×10^3 kg m^{-3} to about 2.5×10^3 kg m^{-3} at 1300°C. Corresponding increases in the fibre elastic modulus and tensile strengths also occur. At temperatures above 850°C the conversion is complete, but the structure is amorphous. Commercially available SiC fibres of this type (Nicalon*) can be considered

---

* Nicalon is a trade name of the Nippon Carbon Co.

to consist of amorphous SiC and SiO_2 with the carbon segregated into clusters having a mean radius of 2.0 nm. The average fibre composition is given by Simon and Bunsell (1984a) as 69% SiC, 24% SiO_2 and 7% C by weight, but wide variations in fibre compositions are observed (see e.g. Andersson and Warren, 1984). The fibre strength falls as the temperature is increased above about 1000°C both in air and in argon and Simon and Bunsell (1984a) also note some reduction in Young's modulus at elevated temperatures. The SiO_2 is distributed throughout the fibre and is considered to be present as a consequence of the initial oxidation stage of fibre manufacture. The fibre strengths are degraded at temperatures in excess of 1200°C and this is associated with the loss of CO from the fibre and β-SiC grain growth. The details of this process are dependent upon the heat treatment environment (Mah et al., 1984).

The relatively low fibre Young's modulus is consistent with the presence of amorphous SiO_2 and carbon particles. The fibre strengths are controlled by a distribution of flaws and are therefore dependent upon the gauge length tested. Measured average strength values range from about 2.5 GN m^{-2} for a 10 mm gauge length (Andersson and Warren, 1984) to 1.3 GN m^{-2} for a 200 mm gauge length (Simon and Bunsell, 1984a). Simon and Busnell (1984b) have investigated the creep characteristics of fibres of this type. Tensile strengths are reduced to very low values at temperatures in excess of 1300°C, so that creep is of interest in the range 1000–1300°C. The creep processes are complex because the fibre microstructure changes with time at a rate which is dependent on the temperature, stress and atmospheric environment. The SiO_2 component of the fibres, which is viscous at high temperatures, would also be expected to enhance the creep deformation of the fibres. Okamura et al. (1983) have pointed out that experimental fibres can be prepared without utilizing an oxidative stabilization step, which is the origin of the SiO_2 content.

2.8.3 SiC Whiskers from Rice Hulls

Rice hulls contain between 15% and 20% by weight of SiO_2. The production of silicon carbide whiskers from this material has been described by Lee and Cutler (1975). The rice hulls are first heated to a temperature of 700–900°C in an oxygen-free atmosphere to remove volatiles; the residue contains approximately equal weights of SiO_2 and free carbon. SiC is formed by heating the residue in a stream of N_2 or NH_3 gas to temperatures between 1500°C and 1600°C for about 1 h. The reaction proceeds via the formation of SiO and the CO formed has to be flushed away by the flowing gas atmosphere in order to maintain the reaction. When the reaction is complete the residue is heated to 800°C in air to remove the remaining free carbon. The resulting yield of SiC is about 10% of the original rice hulls. Small proportions of SiO_2 and Si_3N_4 are also present.

Most of the SiC is present in the form of particles, only about 10% being in the form of whisker crystals. These have diameters of 0.1–1 μm and lengths of about 50 μm. Separation techniques are used to reduce the particle content to about 10%. Divecha *et al.* (1981) quote tensile strengths of 13 GN m^{-2} and Young's modulus values of 700 GN m^{-2} and have described the application of SiC whiskers to the reinforcement of aluminium.

2.8.4   SiC Whiskers Produced by the VLS Process

A liquid catalyst is used in the VLS process. Crystal growth occurs by the transport of material in vapour form to the surface of a molten droplet of the catalyst. The feed material is then deposited from the molten droplet to form a single-crystal whisker that increases in length as material is transported from the vapour through the liquid to the solid crystalline phase. Hence the term VLS process. A substrate is initially seeded with particles of a suitable catalyst—iron alloys can be used for SiC—and under suitable conditions crystal growth occurs. The technique is illustrated schematically in Figure 2.23.

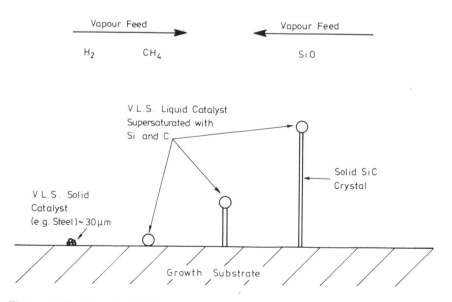

Figure 2.23. Growth of SiC whiskers by the VLS process (schematic). (Redrawn from Milewski *et al.*, 1985.)

A production process based on this technique has been described by Milewski *et al.* (1985). A graphite reactor is used and the required silicon monoxide is generated by reacting a fine powder mixture of silica and carbon according to the reaction

$$SiO_2 + C \rightarrow SiO + CO$$

and CO is added to the reactor feed gases to control this reaction. Other feed gases include CH_4, H_2 and N_2.

The form of the SiC whiskers produced can be controlled by the composition of the gas in the reactor. Under appropriate conditions whisker crystals about 6 μm in diameter and 5–100 mm long can be manufactured. They have a rounded triangular cross-section, consist of β-SiC and grow in the [111] crystallographic direction.

The tensile strengths and elastic moduli of these fibres have been described by Petrovic *et al.* (1985). Tensile strengths for a 5 mm gauge length are reported to be variable, with average values of 8.4 GN m^{-2}. The Young's modulus of the fibre is observed to be about 580 GN m^{-2}, which corresponds reasonably well with the calculated elastic modulus of β-SiC in the [111] direction.

2.9 KEVLAR ARAMID FIBRES

Organic fibres having the trade name Kevlar have been produced by the DuPont Company. These fibres have tensile elastic modulus values up to about 125 GN m^{-2}, tensile strengths approaching 3 GN m^{-2} and densities of about 1.45×10^3 kg m^{-3}. The processes used in their manufacture have been outlined by Magat (1980).

Fibres can be spun from aromatic polyamide liquid crystalline solutions and the freshly spun fibres undergo recrystallization during short periods of heat treatment at temperatures in excess of 350°C. This produces an increase in the tensile elastic modulus and strength of the fibres. Thus, by varying the post-spinning heat treatment, the fibre mechanical properties can be varied over a wide range to suit various engineering applications.

The molecular structure of an aromatic polyamide fibre, Kevlar 49, has been described by Dobb *et al.* (1979). These fibres are formed from rod-like molecules of poly-*p*-phenylene terephthalamide. The fibres are characterized by —CONH— links in the *para* position between aromatic rings, giving a fairly rigid chain together with a large number of hydrogen bonds per unit volume between CO and NH on adjacent chains. The molecules form slightly pleated extended sheets that are aligned longitudinally with the fibre axis and organized in a radial array.

The mechanical properties of the fibre can be associated directly with its structural features. Both the tensile elastic modulus and tensile strength are high. The shear modulus is low at $1.8\,\mathrm{GN\,m^{-2}}$ in torsion, the poor lateral cohesion reflecting the weak bonding between the sheets. The same feature can account for the low compressive strengths, which would be expected to result from a delamination mechanism.

The mechanics of compressive failure of Kevlar 49 fibres has been studied recently by Deteresa *et al.* (1984). Kink bands are observed to develop at compressive strains of about 0.5% and this corresponds approximately to the failing strain in compression of unidirectionally reinforced Kevlar 49 composites. Thus the fibre and composite compressive strengths are only about one-fifth of the corresponding tensile values.

The kink bands extend helically around the fibre, and apparently through the fibre cross-section, with right and left-handed pitch angles of 50–60°. At compressive strains of about 3% the fibre segments between the kink bands become displaced with respect to the fibre axis. During subsequent tensile loading the kink bands developed in compression unfold and appear only as stretch marks or depressions on the fibre surface. The tensile strength of fibres previously stressed in compression so as to produce kink bands is little different from that of undamaged fibres. After the kink bands have unfolded, the fibre elastic modulus, in tension, is restored to that of undamaged fibres.

The low density of the fibre coupled with the very high fibre tensile strength gives a strength-to-density ratio greater than that of other commercially available reinforcing fibres. The elastic modulus-to-density ratio, however, is lower than for high-elastic-modulus graphitic fibres. Room-temperature properties are not significantly degraded at temperatures up to about 180°C, and the chemical resistance of the material is excellent except to a few strong acids and alkalis. However, exposure to sunlight does degrade the fibre properties. The mechanical properties of Kevlar 49–epoxy composites are reduced in a similar manner to those of glass fibre and graphitic fibre analogues under hot and humid conditions. Kevlar aramid fibre is also manufactured in a lower elastic modulus form (Kevlar 29) that has a tensile strength comparable with Kevlar 49 but an elastic modulus about one-half as great.

Kevlar 49 fibres are produced in continuous-filament yarns, the individual fibres having a circular cross-section and a diameter of about $10\,\mu\mathrm{m}$. Kevlar 29 fibres have the same diameter and this material is also available in the form of staple fibres and non-woven felts. The properties of this material are of value in the manufacture of polymeric matrix composite materials having a high resistance to ballistic impact. Both Kevlar 49 and Kevlar 29 are available as fabrics, conventional textile fabrication processes being used for this purpose. These products are similar to glass fibre fabrics and facilitate the manufacture of composite structures using wet lay-up techniques.

2.10 POLYCRYSTALLINE OXIDE FIBRES

Various processes have been developed by which polycrystalline fibres of various oxides can be manufactured. These have been described in some detail by Kelsey (1967). The dimensions of the individual crystals are very small by comparison with the diameters of the fibres, which have a very low porosity, a random orientation of the crystal grains and tensile strengths considerably in excess of that of the corresponding bulk material. Fibres are manufactured by first producing a liquid containing the material that will eventually form the fibre. This can take the form of a suspension or a colloidal solution of a metal organic compound from which the metal ion of the fibre is to be obtained. The precursor fibre is then produced by extruding the liquid through an orifice. Alternatively, the solution can be spread on a flat surface so that, by controlled evaporation of the solvent, multiple cracking of the residual solid occurs with the formation of precursor fibres. The remaining organic material is then removed and polycrystalline oxide fibres are produced by further heat treatment in air.

The properties of polycrystalline Al_2O_3 fibres manufactured by the liquid extrusion process have been described by Dhingra (1980). An aqueous slurry, with various components added to produce appropriate viscosity characteristics, is extruded through orifices to form a precursor spun yarn. This is then heat treated to form a low-porosity polycrystalline Al_2O_3 fibre yarn. The individual fibres are then coated with a thin silica layer, the heat treatment associated with this step probably forming an aluminosilicate layer. This process increases the effective fibre strength by reducing the severity of the initial surface flaws. Tensile strengths approaching $2 \, GN \, m^{-2}$ are achievable with elastic moduli of $380 \, GN \, m^{-2}$. The fibre density, however, is relatively high ($3.9 \times 10^3 \, kg \, m^{-3}$) compared with other reinforcing fibres such as carbon.

2.11 DISCONTINUOUS REINFORCING FIBRES

2.11.1 Composite Fabrication using Continuous and Discontinuous Reinforcing Fibres

Certain types of composite fabrication techniques are only possible if continuous fibres are available. Such techniques include filament winding and pultrusion processes. In addition, the manufacture of pre-impregnated warp sheet and of woven textile reinforcement is facilitated by the availability of continuous fibres. Discontinuous fibres can be used effectively to reinforce a matrix, providing the short fibres have a sufficiently high length-to-diameter ratio to provide adequate load transfer between fibres and matrix. The mechanics of this process are discussed in §§3.11 and 5.1. Before the advent of boron fibres and subsequently of carbon fibres, the only high-elastic-modulus high-strength low-density fibres available were whisker crystals, which can only be produced as discontinuous fibres. Composite fabrication processes utilizing discontinuous fibres were therefore developed.

In order to obtain effective reinforcement from discontinuous fibres it is necessary for them to be present at fairly high volume fractions and, to achieve this, their alignment in the composite structure has to be controlled. Whisker crystals grown from the vapour phase are produced in the form of a very-low-density felt and contain crystalline particles and powder. In order for them to be useful as reinforcing fibres it is necessary for them to be separated, cleaned, graded for size and, if possible, to be aligned in a unidirectional array. Techniques by which these requirements could be met have been described by Parratt (1966). Large amounts of carrier liquids are used: stirring the mixture produces velocity gradients in the liquid that dismantle the agglomeration of whiskers. Lumps of grit and coarse whiskers can be removed by elutriation columns and the separation of whiskers into various lengths, together with the removal of micrometre-sized dust particles, is achieved using rotary filter sieves from which the whiskers are continuously removed. The practical achievable volume fraction of whiskers in a random planar felt is quite low. This can be increased considerably if the whiskers are aligned. Processes based on the extrusion of whisker crystals in suspension in a viscous liquid were developed for this purpose.

The techniques originally devised to produce an aligned array of whisker crystals have been developed subsequently to produce aligned felts from continuous fibres that have been chopped into short lengths. The sheet material produced in this way is impregnated with a suitable quantity of polymeric matrix material, which is partly cured, and the material is termed a prepreg. The felt retains its overall shape when uniformly supported but can be deformed locally by external loads that produce relative longitudinal displacements between the short fibres. This has certain advantages over continuous fibres when a composite structure of complex shape is to be reinforced.

When a cloth is draped over a curved surface the angle between the warp and the weft adjusts itself so that the curvature of the fabric fits the surface. Hence the final orientation of the fibres forming the cloth is determined by the curvature of the surface, not by the stress system. The prepreg formed from discontinuous fibre sheets, stacked with the fibres in the individual sheets aligned in particular directions, can be deformed in a similar fashion, but with much less disturbance to the alignment of the discontinuous fibres in the individual plies. The technique possesses a further useful feature in that mixtures of short fibres forming hybrid systems can be dispersed uniformly and arranged in an aligned felt.

2.11.2 Production of Aligned Felts from Discontinuous Fibres

The discontinuous fibres are carried as a dilute suspension in glycerol and extruded through a convergent nozzle to produce a preferred alignment. The fibres then impinge on a porous surface that is moving at right angles to the direction of the flow through the nozzle at a speed slightly greater than the flow from the nozzle (Dingle, 1974). The carrier liquid is removed quickly so that the fibres become aligned on the porous surface in the direction of the relative motion between it and the flow of the dispersion through convergent nozzle.

Recent developments of this technique have been described by Edwards and Evans (1980). Initially the fibre mat was formed by extruding the suspension of fibres through a slit that was reciprocated over a filter bed, the carrier liquid being removed by suction. Improved fibre alignment is obtained by extruding the fibre suspension through a tapering nozzle instead of a slit. The mat is formed by reciprocating the jet over the inner surface of a permeable cylinder rotating about a horizontal axis and moving with an angular velocity sufficiently great to pass the carrier liquid rapidly through the porous surface by centrifugal force. In the system described by Edwards and Evans the cylinder of the centrifuge consists of perforated stainless steel covered with a 100-mesh nylon gauze for convenient removal of the aligned fibre felt.

2.12 OTHER REINFORCING FIBRES

It is impracticable to deal adequately, within a single chapter, with all of the various types of reinforcing fibres that have been studied and reported on over the last 25 years. A few additional examples are given here.

Single-crystal Al_2O_3 rods (Mehan and Noone, 1974) and refractory metal wires (Signorelli, 1974) have been considered as reinforcing members for use with high-temperature metal alloys. High-strength steel wires have been used to reinforce a polymeric matrix and beryllium wire has received attention because of its high elastic modulus and low density (Roberts, 1967). As mentioned in §1.3.2, refractory reinforcing fibres can be grown within a metal matrix by the controlled solidification of suitable metal alloys. Extensive work has been carried out on this topic, which is beyond the scope of this book. Reinforcing fibres can also be produced by reacting a precursor fibre with a molten matrix. The production of silicon carbide fibres by infiltrating carbon fibres with molten silicon provides an example. In this system the matrix is formed from excess silicon (Mehan, 1978). Various other refractory materials such as boron nitride, silicon nitride, boron carbide and titanium diboride have also been prepared in fibre form.

3
Elastic Properties of Fibrous Composites

Fibre-reinforced composites utilize a wide range of materials and are fabricated in various ways. The matrix materials may be polymeric ranging from rubber to brittle resins, ceramic materials and also metals. Various types of reinforcing fibres may be incorporated in the same composite and the fibres can be present in various volume fractions and be orientated randomly or be aligned in various ways. The non-homogeneous micro-structure causes highly anisotropic elastic properties to be developed in composites containing fibres that are orientated preferentially in specific directions. Some of the properties of the composite are influenced by the absolute size of the fibres which may, or may not, be of circular cross-section. The fibres may function as individual reinforcing members; they may be agglomerated to form bundles; or they may be arranged as laminae of uniform thickness within which the individual fibres are unidirectionally aligned. The properties of the laminated system are then controlled by the fibre alignment in the various laminations. Although they are primarily used as stiff materials, fibrous composites may also be used in the form of flexible structures and their electrical and magnetic properties are also important in some applications. The fibres are usually much stiffer than the matrix, but the reverse is true in some cases. However the fibres are always stronger than the unreinforced matrix. Generally, external loads applied to the matrix are redistributed via the fibres to the rest of the composite structure.

Practical fibrous composites have properties that are very variable on the microscopic scale. In general the fibres will not be fully continuous and will contain flaws varying in severity and spatial distribution. The distances separating adjacent fibres in a matrix are quite variable. The matrix may contain voids or flaws and the fibre–matrix interface may have non-uniform properties. However, despite the problems inherent in a detailed characterization of fibrous composites, it is possible to deal analytically with a number of important issues through an examination of idealized systems.

In §3.1, we first review briefly and at an elementary level the relationships between the elastic properties of an ideal isotropic elastic solid before examining in later sections the factors that control the anisotropic elastic properties of fibre-reinforced composites.

3.1 RELATIONSHIPS BETWEEN THE ELASTIC CONSTANTS OF AN ISOTROPIC ELASTIC MATERIAL

We neglect any permanent deformation that may be caused in a body by the application of a force and consider only reversible elastic conditions under which a body returns to its original shape after the removal of an externally applied force. We define strain as the change in shape or size produced by the application of external forces and stress as the system of forces in equilibrium that are required to produce and maintain the strain. We also assume that the strains are small, that the applied stress is directly proportional to the strain developed, and that the strains developed by various systems of forces are additive.

We consider first the situation in which a body is subjected to a unidirectional tensile stress as indicated in Figure 3.1. The stress σ is given by the applied force P divided by the cross-sectional area A, and the strain ε is defined as the change in length per unit length. The constant of proportionality is the Young's Modulus E, so that $E = \sigma/\varepsilon$. Both E and σ are expressed in terms of Nm^{-2}, so that ε is dimensionless. The development of a tensile strain causes a reduction in the cross-sectional dimensions of the body. The constant of proportionality v is termed Poisson's ratio and is defined as the proportional contraction per unit width divided by the proportional extension per unit length.

If a body (for example a cube of unit dimension, as shown in Figure 3.2) is subjected to a hydrostatic pressure, so that the stresses applied to all faces are equal, its volume will decrease. The constant of proportionality is now defined as the bulk modulus K and is given by the normal stress per unit area of surface divided by the change in volume per unit volume. The applied stresses can of course be tensile, in which case the volume of the material increases under load.

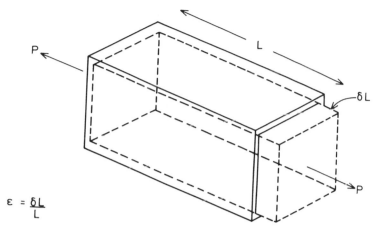

$$\varepsilon = \frac{\delta L}{L}$$

Figure 3.1. Effect of uniaxial tension.

In Figure 3.3, the rectangular block of material is shown deformed under the action of a shear force. The shear stress is defined in terms of the tangential force divided by the area over which it is acting and is usually termed τ. This stress produces an angular deformation γ and the form of deformation illustrated is termed simple shear. The constant of proportionality is G so that $G = \tau/\gamma$ and γ is defined as the engineering shear

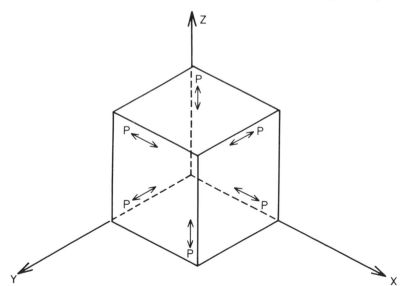

Figure 3.2. Cube subjected to compressive or tensile forces acting in the x,y,z, coordinate system.

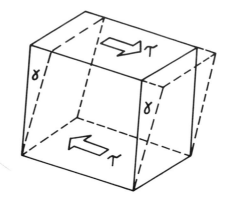

Figure 3.3. Simple shear.

strain. Shear can be considered as equivalent to an extension and contraction at right angles to each other as can be seen from the following simplified argument. In Figure 3.4 the diagonal BD is increased to B′D and the diagonal AC is contracted to A′C by the shear deformation. The proportional extension of BD is given by $(B'D - BD)/BD \sim B'E/BD$ (where BE is perpendicular to DB′). Since the deformation is small this can be written as

$$\frac{BB'/\sqrt{2}}{BC\sqrt{2}} = \frac{1}{2}\frac{BB'}{BC} = \frac{\gamma}{2}$$

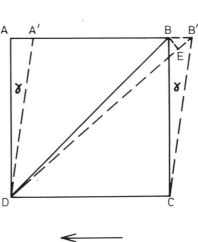

Figure 3.4. Angular shear related to tensile and compressive strains.

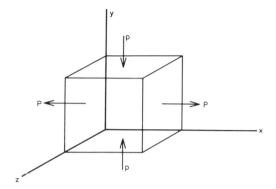

Figure 3.5. Illustrating relationship between E, G and v.

Similarly the proportional contraction along AC is also given by $\gamma/2$. Thus the extension and contraction are each equal to half the angle of shear.

Various relationships can be established between these elastic constants. We first consider a cube of unit dimensions subjected to a compressive force P acting in the y direction on opposite faces of the cube, and subjected also to a tensile force P acting in the x direction. No external forces are assumed to be acting in the z direction (Figure 3.5). The extension strain in the x direction due to the stresses acting in the x direction is given by $+P/E$ and the extension strain in the x direction due to the compressive force acting in the y direction is $+vP/E$. Hence the total tensile strain is $(P/E)(1 + v)$. Similarly, the compressive strain in the y direction is $(P/E)(1 + v)$. There is zero net strain in the z direction since the force acting in the x direction produces a contraction strain of $-vP/E$ in the z direction and the force acting in the y direction produces an extension strain $+vP/E$ in the z direction.

From our previous definition of shear as being equivalent to a tensile and a compressive strain acting at right angles to each other, each equal to half the angle of shear, we have,

$$\frac{P}{E}(1 + v) = \frac{\gamma}{2}$$

The angular deformation γ could also be produced by a tangential force P acting on opposite faces of the unit cube so that,

$$2\frac{P}{E}(1 + v) = \frac{P}{G}$$

and
$$E = 2G(1 + v)$$

We now consider a unit cube subjected to a normal tensile force P acting on all faces, as in Figure 3.2. The extension in the x direction due to the force P acting in the x direction is P/E. Contractions in the x direction each equal to vP/E are produced by the forces P acting in the y and z directions, so that the resultant extension in each direction is given by $(P/E)(1 - 2v)$ and the increase in volume per unit volume can be taken as $(3P/E)(1 - 2v)$. Since the cube is effectively subjected to a hydrostatic tension P we have $(3P/E)(1 - 2v) = P/K$, where K is the bulk modulus of the material. Hence $E = 3K(1 - 2v)$.

Since E and K are both positive it follows that for an isotropic elastic solid v cannot be greater than 0.5, and this corresponds to a situation in which there is no volume change during deformation. These equations can be transposed to give the following relationships.

$$E = \frac{9KG}{3K + G} \quad \text{and} \quad v = \frac{3K - 2G}{2(3K + G)}$$

Fibrous composites are often constructed from several layers, plies or laminae in the form of thin sheets within which the fibres are uni-directionally aligned. For these conditions, stresses normal to the surface of the sheet can be assumed to be zero so that the loading condition corresponds to that of plane stress. This is illustrated in Figure 3.5 where the stresses in the z direction are taken to be zero. For this situation, the stresses acting on the faces of the cube are the normal stresses σ_x, σ_y acting respectively on the 'x' and 'y' surfaces of the cube and τ_{xy}, the shear stress, producing a shear strain γ_{xy} in the xy plane. The strain components ε_x, ε_y and γ_{xy} can be written in terms of the stress components. If we define tensile stresses as positive and compressive stresses as negative, we have,

$$E\varepsilon_x = \sigma_x - v\sigma_y$$
$$E\varepsilon_y = \sigma_y - v\sigma_x \tag{3.1}$$
$$G\gamma_{xy} = \tau_{xy} = \frac{E\gamma_{xy}}{2(1 + v)} = \frac{E}{(1 + v)}\varepsilon_{xy}$$

By transposition,

$$\sigma_x = \frac{E}{(1 - v^2)}(\varepsilon_x + v\varepsilon_y)$$
$$\sigma_y = \frac{E}{(1 - v^2)}(\varepsilon_y + v\varepsilon_x) \tag{3.2}$$
$$\tau_{xy} = G\gamma_{xy}$$

These equations can be written in matrix notation as:

$$\begin{bmatrix} \sigma_x \\ \sigma_y \\ \tau_{xy} \end{bmatrix} = \begin{bmatrix} E/(1-v^2) & vE/(1-v^2) & 0 \\ vE/(1-v^2) & E/(1-v^2) & 0 \\ 0 & 0 & G \end{bmatrix} \begin{bmatrix} \varepsilon_x \\ \varepsilon_y \\ \gamma_{xy} \end{bmatrix} \qquad (3.3)$$

The same notation is used to describe the constitutive relationships for anisotropic elastic materials, but for these substances the terms shown as zero in the above matrix can have finite values.

3.2 DEFORMATIONS AND STRESSES DUE TO TENSILE LOADS APPLIED IN THE DIRECTION OF FIBRE ALIGNMENT IN A UNIDIRECTIONAL SYSTEM

We first consider that both fibres and matrix behave elastically and remain intact. If we neglect any distortions occurring at the ends of the composite, where the external load is being applied, we can assume that both fibres and matrix are undergoing the same tensile strain ε_c the composite tensile strain. Hence, if the Poisson ratios of fibres and matrix are the same and the Young's modulus of the composite is E_c we have,

$$E_c \varepsilon_c = E_f A_f \varepsilon_c + E_m A_m \varepsilon_c \qquad (3.4)$$

where E_f and E_m are respectively the Young's modulus of fibres and matrix and A_f and A_m their relative cross-sectional areas. Equation (3.4) states that the load applied to a unit cross-section of the composite at a particular strain value is supported partly by the fibres and partly by the matrix. Since we assume that the fibres are uniform cylinders uniformly distributed we have, within the elastic range of both materials,

$$E_c = E_f V_f + E_m V_m \qquad (3.5)$$

where V_f and V_m are the volume fractions of fibres and matrix respectively. Equation (3.5), which represents a simple law of mixtures, gives the lower bound for the elastic modulus E_c of the composite. If the Poisson ratios of the fibres and matrix are not equal, the mutual constraints they impose on each other will increase the elastic distortional energy at any particular strain, thus increasing the effective Young's modulus of the composite material (Hill, 1964; Hashin, 1970). If v_f and v_m represent the Poisson ratios of fibres and matrix and G_f and G_m their respective shear moduli, it can be shown that the increase in Young's modulus of the composite over that given by the law of mixtures lies between the limits

$$4V_f V_m (v_f - v_m)^2 [V_f/K_{pm} + V_m/K_{pf} + 1/G_m]^{-1}$$

and

$$4V_f V_m (\nu_f - \nu_m)^2 [V_f/K_{pm} + V_m/K_{pf} + 1/G_f]^{-1}$$

where K_{pm} and K_{pf} are the plane strain bulk moduli of matrix and fibre respectively. These limits are exact solutions for a cylindrical elastic body in which one component is surrounded by a coaxial cylindrical shell of the other component. The former limit corresponds to a fibre encapsulated in a cylindrical matrix sheath. In most practical systems these correction factors only amount to about 1% or 2% of the value obtained from the simple rule of mixtures (equation (3.5)), which can thus be used generally to estimate the value of the Young's modulus of the composite measured parallel to the fibres to within the expected range of experimental error (see Cooper, 1971.)

The composite is usually subjected to temperatures above ambient during fabrication. This occurs, for example, where a thermosetting polymer is used as a matrix. Since, generally, the fibres and matrix will have different thermal expansion coefficients, residual tensile or compressive stresses will be developed at the fibre–matrix interface on cooling to room temperature. In the case of glass-fibre–polymeric-matrix composites the polymer matrix has the higher thermal expansion coefficient, so that at first sight the fibre–matrix interface would be expected to be subjected to compressive stresses during thermal contraction of the matrix. Although this is generally the case when the volume fraction of fibres is low, the stresses developed are not uniform over the fibre surface because of interactions with adjacent fibres. The degree of non-uniformity increases with increasing fibre volume fraction and can result in the development of tensile stresses at the fibre–matrix interface at high volume fractions of fibre (see Owen, 1974). The interfacial tensile stresses developed in this way have maximum values at the positions corresponding to the maximum separation of the fibre surfaces, as can be seen intuitively if the limiting case is considered in which three cylindrical fibres are almost in contact, thus surrounding an approximately triangular region of matrix subjected to thermal contraction. In addition to the stresses developed at the fibre–matrix interface due to temperature changes, similar stresses are developed when a tensile load is applied parallel to the fibre alignment, as a consequence of any difference in Poisson ratios between fibres and matrix, as discussed above. Again these stresses will be influenced strongly by the interfibre spacing. Since, in practical composites, the fibres are distributed non-uniformly, effects due to interactions between adjacent fibres will be pronounced in local regions of high fibre concentration. Longitudinal stresses are also produced by differential thermal contraction of fibres and matrix. For example, the longitudinal thermal expansion of carbon fibres is almost zero so that they are placed in longitudinal compression by the thermal contraction of a polymeric matrix after fabrication.

If the matrix is plastically deformable, as is the case with a metal for example, then the stress–strain curve of the composite will have two regions. Initially both fibres and matrix will deform elastically so that the elastic modulus of the composite is given with reasonable accuracy by equation (3.5). Eventually the matrix will yield causing a change in the slope of the stress–strain curve. The matrix may yield at a strain lower than its intrinsic yield strain owing to internal stresses developed in the composite. Also, if the fibres are very close together ($<10 \, \mu m$) they can interfere with the movement of dislocations in a metallic matrix, so that its properties are changed compared to those with the bulk material. Generally, the slope of the stress–strain curve during the second stage of deformation is very close to $E_f V_f$, since the contribution by the plastically deforming matrix to the elastic properties of the composite is negligible.

If the fibres themselves deform plastically before failing, a further stage of deformation may be observed. This is concluded by ductile failure of the reinforcing fibres and is governed by the onset of plastic instability in the system (Mileiko, 1969).

3.3 GENERAL ELASTIC PROPERTIES OF UNIDIRECTIONALLY REINFORCED FIBROUS COMPOSITES

The idealized composite structure illustrated in Figure 3.6 is orthotropic, having three mutually perpendicular planes of symmetry. Its elastic characteristics are different in different directions. The direction of fibre alignment is termed the 1 direction and mutually perpendicular directions, each perpendicular to the direction of fibre alignment, are termed the 2 and 3 directions. Many composite structures are manufactured from layers of thin unidirectionally reinforced sheets termed laminae or plies.

In the sections that follow, an analysis of the anisotropic elastic characteristics of a unidirectionally reinforced lamina is outlined. Also the methods used by which the equations describing the elastic characteristics of an assembly of laminae bonded together to form a laminate are given. Full descriptions of these analyses with rigorous developments of the mathematical techniques are given in various textbooks on the subject, for example Jones (1975), Ashton et al. (1969) and Vinson and Chou (1975).

We assume that the stresses at a point can be represented by the stresses acting at the surfaces of a cube located at that point. The stresses shown acting on the cube are indicated in Figure 3.7. There are three normal stresses (σ_{11}, σ_{22} and σ_{33}) and six shear stresses (τ_{31}, τ_{21}, τ_{23}, τ_{32}, τ_{13},

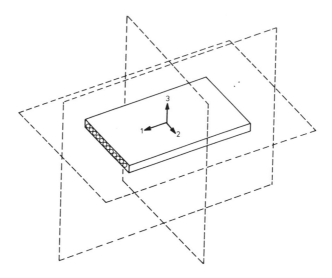

Figure 3.6. Three mutually perpendicular planes of symmetry in a unidirectional reinforced lamina.

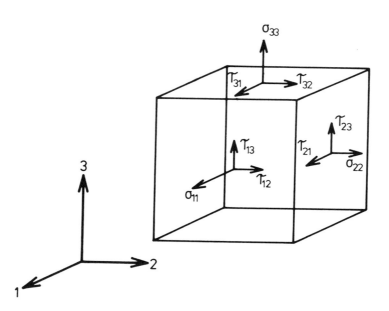

Figure 3.7. Three-dimensional state of stress.

τ_{12}). These are taken as being due to the stresses exerted on the faces of the cube by the material outside the cube. The first suffix refers to the direction normal to the plane in which the stress is acting and the second suffix to the direction in which the stress is acting. Since, for equilibrium, $\tau_{23} = \tau_{32}$, $\tau_{13} = \tau_{31}$ and $\tau_{12} = \tau_{21}$, three shear stresses, τ_{23}, τ_{31} and τ_{12} can be used to represent the shear stresses acting on the cube. Corresponding strains, ε_{11}, ε_{22}, ε_{33}, γ_{23}, γ_{31} and γ_{12} are developed by the normal stresses and shear stresses. A contracted stress notation of σ_1, σ_2 and τ_{12} with corresponding strains of ε_1, ε_2 and γ_{12} is normally used to describe the stresses and associated strains for a lamina under plain stress conditions.

The transverse elastic Young's modulus measured in a direction perpendicular to the fibre alignment is termed E_2 in this notation. By symmetry, from Figure 3.6, $E_2 = E_3$. The in-plane shear modulus in the 1–2 plane is given by G_{12} and E_1 is the Young's modulus measured in the direction of fibre alignment.

3.4   TRANSVERSE ELASTIC MODULUS

The deformation resulting from a tensile stress σ_2 applied in a direction perpendicular to the direction of the fibre alignment can be estimated by assuming that both fibres and matrix are lumped together in a sandwich structure as illustrated in Figure 3.8 and subjected to the same stress. The

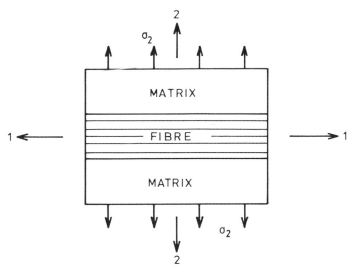

Figure 3.8. Cross-section of representative volume element loaded in the 2 direction.

composite strain ε_2 thus given by

$$\varepsilon_2 = V_f \varepsilon_f + V_m \varepsilon_m$$

where ε_f is the strain in the fibres and ε_m the strain in the matrix. Since the transverse elastic modulus $E_2 = \sigma_2/\varepsilon_2$, we can write

$$\frac{1}{E_2} = \frac{V_f}{E_f} + \frac{V_m}{E_m} \qquad (3.6)$$

Note that here E_f refers to the transverse Young's modulus of the fibres. This will be the same as the axial Young's modulus in the case of a homogeneous fibre such as glass, but will not be so in the case of anisotropic fibres such as carbon. However, generally, $E_f \gg E_m$ so that E_2 is given approximately by E_m/V_m.

A more accurate estimate of the transverse Young's modulus can be made using an equation developed by Halpin and Tsai (1969). This equation can be used to calculate other elastic constants and has the form

$$\frac{\bar{P}}{P_m} = \frac{(1 + \zeta \eta V_f)}{(1 - \eta V_f)} \qquad (3.7)$$

where

$$\eta = \frac{(P_f/P_m) - 1}{(P_f/P_m) + \zeta}$$

$\bar{P}$ corresponds to E_2 for fibres in a square array, when ζ is given a numerical value of 2. The corresponding fibre and matrix elastic moduli are substituted for P_f and P_m. Equation (3.7) can also be used to predict values of E_2 for composites reinforced with rectangular-sectioned fibres or tapes. Here ζ is taken as $2a/b$ where a is the width and b is the thickness of the reinforcing members. If a suitable choice is made for the value of the factor ζ, equation (3.7) can be used to predict values in close agreement with more exact computations using formal elasticity theory, except for very high values of V_f. The general applicability of equation (3.7) has been discussed by Jones (1975).

Since the fibres and matrix have different elastic moduli it is apparent that the strains and corresponding stresses, developed under a transverse tensile load, will not be uniform. The maximum stresses occur at the fibre–matrix interface and, for fibres in square array, occurs at the points where the distance between adjacent fibres is a minimum. This local stress concentration depends on the ratio of the elastic modulus of fibres and matrix and the volume fraction of fibre. Adams and Doner (1967b), have calculated the values of maximum stress enhancement as a function of fibre volume fraction and the ratio E_f/E_m. For typical fibre composites the magnitude of the local stress concentration amounts to about a factor of 2. Adams and Doner (1967b) also calculate values of E_2 for various volume fractions of fibre and various ratios of elastic modulus of fibre to that of matrix.

3.5 LONGITUDINAL SHEAR LOADING

An estimate of the longitudinal shear modulus of a unidirectional composite can be obtained by assuming that both fibres and matrix are subjected to the same shear stress. This condition is illustrated in Figure 3.9. Again the fibres are assumed to be agglomerated to form the central layer of a 3-layer structure. We assume that the shear deformation in the matrix is given by $\gamma_m = \tau/G_m$ and the shear deformation in the fibres is given by $\gamma_f = \tau/G_f$. These deformations are indicated in Figure 3.9. The total shear deformation of the assembly is taken as $\Delta = \gamma W$ where γ is the average shear deformation and W the thickness of the assembly. We have $\Delta = \Delta_m + \Delta_f = V_m W \gamma_m + V_f W \gamma_f$, so that

$$\gamma = V_m \gamma_m + V_f \gamma_f$$

and by substituting G_{12}, G_m and G_f for γ, γ_m and γ_f respectively we have

$$\frac{1}{G_{12}} = \frac{V_f}{G_f} + \frac{V_m}{G_m} \tag{3.8}$$

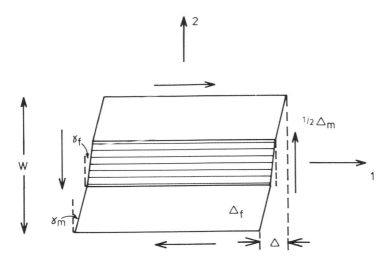

Figure 3.9. Cross-section of representative volume element loaded in shear. (After Jones, 1975.)

This is the same type of equation as the expression (equation (3.6)) derived for E_2 and again shows the matrix elastic modulus as the dominant term.

The Halpin–Tsai relationship (equation (3.7)) can be used to estimate the value of G_{12} to a rather higher degree of accuracy than can equation (3.8), providing that a suitable value is used for ζ. Very good agreement with the calculations carried out by Adams and Doner (1967a), using formal elasticity theory, for circular-sectional fibres at a volume fraction of 0.55, is obtained by taking $\zeta = 1.0$.

Local regions of high shear stress are developed at the fibre–matrix interface. The maximum values occur at the points of closest proximity between adjacent filaments, the effect being similar to the stress concentrations developed under transverse tensile loading and becoming very large when the fibre separation distance is small.

3.6 POISSON RATIOS

If a unidirectionally aligned composite is loaded in tension in the direction of the fibre alignment, both the fibres and the matrix will contract in a direction normal to the stress direction as a consequence of Poisson contradiction. For this condition we can again, as first approximation, lump the fibres and matrix together in a sandwich structure. The concentration due to the fibres in the 2 direction due to a strain ε_1 in the 1 direction will be given by $-\varepsilon_1 V_f \nu_f$, where ν_f is the Poisson ratio of the fibres. Similarly, the matrix contraction will be given by $\varepsilon_1 V_m \nu_m$. Hence the Poisson contraction for the composite structure ν_{12} will be given by a simple rule of mixtures relationship:

$$\nu_{12} = V_f \nu_f + V_m \nu_m \qquad (3.9)$$

By symmetry $\nu_{12} = \nu_{13}$. Note that the Poisson contraction in the direction of fibre alignment (the 1 direction) due to tensile stresses applied in the 2 or 3 directions cannot be estimated in this way. This has been pointed out very clearly by Jones (1975), and the situation is illustrated in Figure 3.10. Taking $^1\Delta_1$ as denoting the displacement in the 1 direction due to a stress σ, we have $^1\Delta_1 = \sigma L/E_1$ and $^1\Delta_2 = (\nu_{12}/E_1)\sigma L$, the direction of loading being denoted by the prefixed superscript. If we apply the same stress in the 2 direction we have $^2\Delta_2 = \sigma L/E_2$ and $^2\Delta_1 = (\nu_{21}/E_2)\sigma L$. Since $E_1 > E_2$ then $^1\Delta_1 < ^2\Delta_2$. Also, from the Reciprocal Theorem developed by Maxwell, Betti and Rayleigh, the appropriate transverse deformation must be the same whether the stress is applied in the 1 direction or the 2 direction so that

$$^1\Delta_2 = {}^2\Delta_1 \qquad \text{and} \qquad \nu_{21}/E_2 = \nu_{12}/E_1 \qquad (3.10)$$

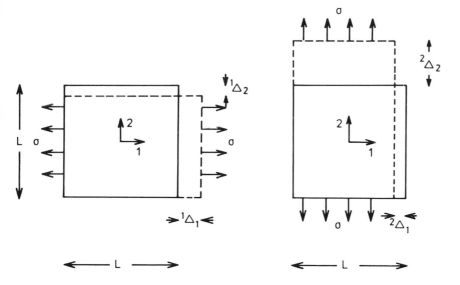

Figure 3.10. Distinction between v_{12} and v_{21}. (After Jones, 1975.)

3.7 OFF-AXIS ELASTIC PROPERTIES OF A UNIDIRECTIONALLY REINFORCED LAMINA

The elastic properties of an isotropic lamina were defined in matrix notation in equation (3.3). A similar set of relationships can be derived for a unidirectionally reinforced lamina loaded in tension or compression along the principal material directions (figure 3.6). Assuming plane stress conditions, the corresponding matrix notation for these relationships is given by,

$$\begin{bmatrix} \sigma_1 \\ \sigma_2 \\ \tau_{12} \end{bmatrix} = \begin{bmatrix} Q_{11} & Q_{12} & 0 \\ Q_{12} & Q_{22} & 0 \\ 0 & 0 & Q_{66} \end{bmatrix} \begin{bmatrix} \varepsilon_1 \\ \varepsilon_2 \\ \gamma_{12} \end{bmatrix} \qquad (3.11)$$

where

$$Q_{11} = E_1/(1 - v_{12}v_{21})$$
$$Q_{12} = \frac{v_{12}E_2}{(1 - v_{12}v_{21})} = \frac{v_{21}E_1}{(1 - v_{12}v_{21})} \qquad (3.12)$$
$$Q_{22} = E_2/(1 - v_{12}v_{21})$$
$$Q_{66} = G_{12}$$

Equation (3.11) reduces to equation (3.3) if $E_1 = E_2$ and $v_{12} = v_{21}$. If the stiffness matrix $[Q]$ is inverted by matrix manipulation the compliance matrix $[S] = [Q]^{-1}$ is obtained.

The corresponding strain–stress relations can then be written,

$$
\begin{bmatrix} \varepsilon_1 \\ \varepsilon_2 \\ \gamma_{12} \end{bmatrix} = \begin{bmatrix} S_{11} & S_{12} & 0 \\ S_{12} & S_{22} & 0 \\ 0 & 0 & S_{66} \end{bmatrix} \begin{bmatrix} \sigma_1 \\ \sigma_2 \\ \tau_{12} \end{bmatrix} \tag{3.13}
$$

where

$$
\begin{aligned}
S_{11} &= 1/E_1, \qquad S_{22} = 1/E_2 \\
S_{12} &= -v_{12}/E_1 = -v_{21}/E_2 \\
S_{66} &= 1/G_{12}
\end{aligned} \tag{3.14}
$$

The Q_{ij} terms in equation (3.11) are called reduced stiffnesses and are related to the terms in the compliance matrix in the following way:

$$
Q_{11} = \frac{S_{22}}{S_{11}S_{22} - S_{12}^2}
$$

$$
Q_{12} = \frac{-S_{12}}{S_{11}S_{22} - S_{12}^2}
$$

$$
Q_{22} = \frac{S_{11}}{S_{11}S_{22} - S_{12}^2}
$$

$$
Q_{66} = 1/S_{66}
$$

When a unidirectionally reinforced orthotropic lamina is loaded only in the principle material directions, the deformation is independent of G_{12} and there is no coupling between the tensile and shear strains. This is not the case if the load is applied in the plane of the laminate at some arbitrary angle to the direction of fibre alignment (Figure 3.11).

We can express stresses in the x, y coordinate system in terms of stresses in the 1, 2 coordinate system in the following way using matrix notation (see e.g. Sokolnikoff, 1956).

$$
\begin{bmatrix} \sigma_x \\ \sigma_y \\ \tau_{xy} \end{bmatrix} = \begin{bmatrix} \cos^2 \theta & \sin^2 \theta & -2 \sin \theta \cos \theta \\ \sin^2 \theta & \cos^2 \theta & 2 \sin \theta \cos \theta \\ \sin \theta \cos \theta & -\sin \theta \cos \theta & \cos^2 \theta - \sin^2 \theta \end{bmatrix} \begin{bmatrix} \sigma_1 \\ \sigma_2 \\ \tau_{12} \end{bmatrix} \tag{3.15}
$$

where θ is the angle measured from the x axis to the 1 axis.

The same matrix can be used to transform the strain values to x, y coordinates if the tensor definition of shear strain is used (i.e. half the "engineering" shear strain). By using the transformation matrix (see e.g.

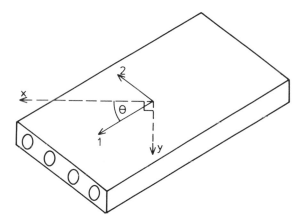

Figure 3.11. Rotation of lamina coordinate system from 1–2 to x–y axes.

Ashton *et al.* 1969), the transformed reduced stiffness matrix $\bar{Q}_{ij}$ is generated thus,

$$\begin{bmatrix} \sigma_x \\ \sigma_y \\ \tau_{xy} \end{bmatrix} = \begin{bmatrix} \bar{Q}_{11} & \bar{Q}_{12} & \bar{Q}_{16} \\ \bar{Q}_{12} & \bar{Q}_{22} & \bar{Q}_{26} \\ \bar{Q}_{16} & \bar{Q}_{26} & \bar{Q}_{66} \end{bmatrix} \begin{bmatrix} \varepsilon_x \\ \varepsilon_y \\ \gamma_{xy} \end{bmatrix} \tag{3.16}$$

where

$$\bar{Q}_{11} = Q_{11}c^4 + 2(Q_{12} + 2Q_{66})s^2c^2 + Q_{22}s^4 \tag{3.17}$$

$$\bar{Q}_{22} = Q_{11}s^4 + 2(Q_{12} + 2Q_{66})s^2c^2 + Q_{22}c^4 \tag{3.18}$$

$$\bar{Q}_{12} = (Q_{11} + Q_{22} - 4Q_{66})s^2c^2 + Q_{12}(s^4 + c^4) \tag{3.19}$$

$$\bar{Q}_{66} = (Q_{11} + Q_{22} - 2Q_{12} - 2Q_{66})s^2c^2 + Q_{66}(s^4 + c^4) \tag{3.20}$$

$$\bar{Q}_{16} = (Q_{11} - Q_{12} - 2Q_{66})sc^3 + (Q_{12} - Q_{22} + 2Q_{66})s^3c \tag{3.21}$$

$$\bar{Q}_{26} = (Q_{11} - Q_{12} - 2Q_{66})s^3c + (Q_{12} - Q_{22} + 2Q_{66})sc^3 \tag{3.22}$$

where $s = \sin\theta$ and $c = \cos\theta$. Notice that when $\theta = 0$, $\bar{Q}_{16}$ and $\bar{Q}_{26}$ become zero and equation (3.16) reduces to equation (3.11). For $0° < \theta < 90°$, shear stress–tensile stress interactions occur.

Using the same matrix manipulations, the transformed reduced compliance matrix $\bar{S}_{ij}$ is generated thus,

$$\begin{bmatrix} \varepsilon_x \\ \varepsilon_y \\ \gamma_{xy} \end{bmatrix} = \begin{bmatrix} \bar{S}_{11} & \bar{S}_{12} & \bar{S}_{16} \\ \bar{S}_{12} & \bar{S}_{22} & \bar{S}_{26} \\ \bar{S}_{16} & \bar{S}_{26} & \bar{S}_{66} \end{bmatrix} \begin{bmatrix} \sigma_x \\ \sigma_y \\ \tau_{xy} \end{bmatrix} \tag{3.23}$$

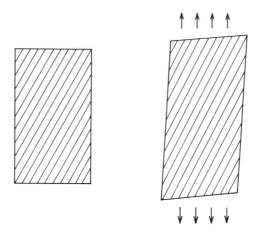

Figure 3.12. Illustrating change in shape of a single lamina subjected to an off-axis tensile load.

where

$$\bar{S}_{11} = S_{11}c^4 + (2S_{12} + S_{66})s^2c^2 + S_{22}s^4 \tag{3.24}$$

$$\bar{S}_{12} = S_{12}(s^4 + c^4) + (S_{11} + S_{22} - S_{66})s^2c^2 \tag{3.25}$$

$$\bar{S}_{22} = S_{11}s^4 + (2S_{12} + S_{66})s^2c^2 + S_{22}c^4 \tag{3.26}$$

$$\bar{S}_{66} = 2(2S_{11} + 2S_{22} - 4S_{12} - S_{66})s^2c^2 + S_{66}(s^4 + c^4) \tag{3.27}$$

$$\bar{S}_{16} = (2S_{11} - 2S_{12} - S_{66})sc^3 - (2S_{22} - 2S_{12} - S_{66})s^3c \tag{3.28}$$

$$\bar{S}_{26} = (2S_{11} - 2S_{12} - S_{66})s^3c - (2S_{22} - 2S_{12} - S_{66})sc^3 \tag{3.29}$$

Again $s = \sin \theta$ and $c = \cos \theta$.

Again equation (3.23) reduces to equation (3.13) when $\theta = 0$. For $0° < \theta < 90°$, shear strain–tensile strain interactions occur and a unidirectional lamina subjected to a tensile strain will deform partly in tension and partly in shear. The effect is illustrated in Figure 3.12.

Using equations (3.14) and (3.24) to (3.29) and taking $\bar{S}_{11}$ as $1/E_x$ etc., the engineering elastic properties in the x, y coordinate system can be related to the engineering elastic properties in the $1, 2$ system so that

$$\frac{1}{E_x} = \frac{c^4}{E_1} + \left(\frac{1}{G_{12}} - \frac{2v_{12}}{E_1}\right)s^2c^2 + \frac{s^4}{E_2} \tag{3.30}$$

$$\frac{1}{E_y} = \frac{s^4}{E_1} + \left(\frac{1}{G_{12}} - \frac{2v_{12}}{E_1}\right)s^2c^2 + \frac{c^4}{E_2} \tag{3.31}$$

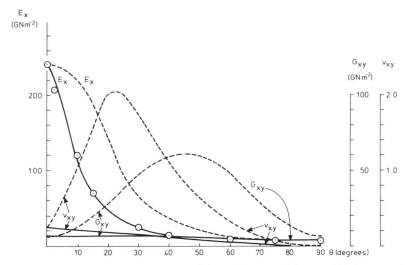

Figure 3.13. Calculated Young's modulus, shear modulus and Poisson ratio as a function of fibre orientation θ. Full lines refer to a single lamina, dashed lines to a $\pm \theta°$ symmetrical laminate (see §3.8.). Material: carbon fibre epoxy resin. Experimental values for the lamina indicated by O. (Data from Sinclair and Chamis, 1979).

$$\frac{1}{G_{xy}} = 2\left(\frac{2}{E_1} + \frac{2}{E_2} + \frac{4v_{12}}{E_1} - \frac{1}{G_{12}}\right)s^2c^2 + \frac{(s^4 + c^4)}{G_{12}} \qquad (3.32)$$

$$v_{xy} = E_x(v_{12}(s^4 + c^4)/E_1 - (1/E_1 + 1/E_2 - 1/G_{12})s^2c^2) \qquad (3.33)$$

Two additional coefficients of mutual influence emerge because of the finite values of $\bar{S}_{16}$ and $\bar{S}_{26}$ in equation (3.23). Values of $E_x E_y G_{xy}$ and v_{xy} are shown as a function of the angle θ for a typical lamina in Figure 3.13 (full lines). Curves of a similar form are obtained for other laminae having different elastic constants.

The validity of equation (3.30) has been confirmed by Sinclair and Chamis (1979), using several tensile specimens cut from a single lamina to cover values of θ from 0° to 90°. The experiment is difficult to perform because an off-axis lamina loaded in tension deforms in the manner indicated in Figure 3.14, and this causes additional deformations to occur in the vicinity of the points of attachment to the loading machine. Sinclair and Chamis obtained stress–strain data from strain gauges mounted at a position mid-way between the loading points and observed a linear elastic response to failure. Their experimental values for E_x are shown in Figure 3.13 and are in close agreement with those computed from equation (3.30) using the experimentally observed lamina properties of 241, 7.72, 0.27, 0.00865 and 6.13 GN m^{-2} for E_1, E_2, v_{12}, v_{21} and G_{12} respectively.

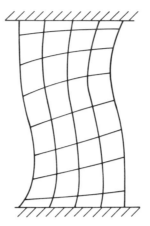

Figure 3.14. Illustrating the deformation of an off-axis lamina under tension and with restrained ends.

3.8 PROPERTIES OF LAMINATES

Because of the severe anisotropy in the elastic properties of unidirectional laminae, these materials are normally used in the form of a laminate. This consists of a stack of laminae, or plies, bonded together by a common matrix. The orientation of the fibres in the individual laminae, the laminae thicknesses and their positions in the laminate then control the elastic properties of the laminate. In the discussion which follows it is assumed that the same type of fibre and matrix and the same volume fraction of fibres are present in all laminae so that all have the same anisotropic elastic properties. The orientation of the fibres within a particular lamina and the thickness of a particular lamina may however be different for laminae located at different positions within the laminate.

It is also assumed that the laminate is thin and that a straight line drawn perpendicular to the mid-plane of the laminate at any point remains straight and perpendicular as the laminate is stretched and bent. This is equivalent to disregarding any shear deformations in planes perpendicular to the plane of the laminate. Strains in the 3 or z direction, i.e. perpendicular to the plane of the laminates, are also ignored. Because of the anisotropy in the elastic characteristics of a lamina, an in-plane tensile load can produce in-plane shear deformation as illustrated in Figure 3.12. If a simple laminate consisting of a $+\theta$ and $-\theta$ ply is subjected to a tensile load, the shear deformations occurring in the individual laminae cause the laminate to twist as illustrated in Figure 3.15. Similarly, in-plane loads can induce bending in addition to in-plane extension. These effects will depend on the

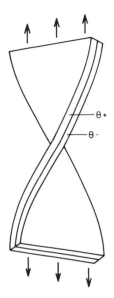

θ+
θ-

Figure 3.15. Twisting induced in unsymmetrical angle-ply laminate loaded in uniaxial tension.

orientation of the fibres in the various plies with respect to the applied loads, to the thickness of each particular lamina, and to its distance from the mid-plane of the laminate.

The polymeric matrix used in fibre-reinforced laminates is generally cured at temperatures appreciably above those at which the laminate will be used. Also, the thermal expansion coefficients are different for different orientations in the laminae. In general, the forces produced by differential thermal contraction produce bending and twisting effects in a laminate and can also produce cracks in some plies. A comprehensive treatment of this subject is beyond the scope of this book but forms part of the general theory of laminates and is discussed by Jones (1975).

Although changes in laminate configuration due to applied loads and thermal effects can be used to advantage in the design of sophisticated structures used in some aerospace engineering applications, they are generally disadvantageous and are usually eliminated by constructing symmetric laminates. For this class of laminates each lamina on one side of the mid-plane of the laminate has its counterpart (in terms of thickness, mechanical properties and fibre orientation) located at the same distance on the other side in the mid-plane (Figure 3.16). It follows that in-plane loads do not produce bending or twisting effects in the laminate because any effect of this nature produced by one particular lamina is balanced by a similar

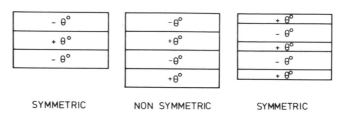

Figure 3.16. Examples of symmetric and non-symmetric angle-ply laminates.

effect produced by its counterpart located at the same distance on the other side of the mid-plane of the laminate.

The constitutive equations describing the in-plane elastic behaviour of a symmetric laminate are arrived at in the following way. For laminates in general the in-plane loads N_x, N_y and N_{xy} are defined in terms of the forces developed per unit width of laminate. They are computed by summing the forces developed in each of the laminae over the whole thickness of the laminate. Since we are dealing here with symmetric laminates, in-plane loads produce only in-plane strains. Hence each lamina is subjected to the same strains as those developed in the mid plane of the laminate, ε_x^0, ε_y^0 and γ_{xy}^0. Thus the stresses carried by each lamina can be computed for these applied strains using the transformed reduced stiffness matrix (equation (3.16)). Each stress term (σ_x, σ_y, τ_{xy}) is multiplied by the corresponding lamina thickness to give the force per unit length developed by each lamina in the plane of the laminate. These forces are then summed for all of the laminae in the laminate. The result can be expressed conveniently by constructing a new matrix, each term of which is the sum of the corresponding terms relating to the individual laminae multiplied by their thicknesses. Thus

$$\begin{bmatrix} N_x \\ N_y \\ N_{xy} \end{bmatrix} = \begin{bmatrix} A_{11} & A_{12} & A_{16} \\ A_{12} & A_{22} & A_{26} \\ A_{16} & A_{26} & A_{66} \end{bmatrix} \begin{bmatrix} \varepsilon_x^0 \\ \varepsilon_y^0 \\ \gamma_{xy}^0 \end{bmatrix} \qquad (3.34)$$

where

$$A_{ij} = \sum_{K=1}^{N} (\bar{Q}_{ij})_k (z_k - z_{k-1}) \qquad (3.35)$$

(see Figure 3.17).

The values of N_x, etc., in equation (3.34) give the sum of the in-plane forces applied to all of the laminae per unit width (length) of laminate that generate the in-plane strains ε_x^0, etc. Thus the in-plane forces have to be divided by the total laminate thickness h in order to obtain the average

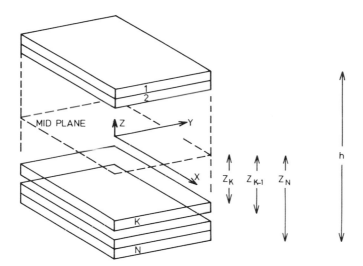

Figure 3.17. Notation for lamina coordinates.

stresses $\bar{\sigma}_x$, $\bar{\sigma}_y$, $\bar{\sigma}_{xy}$ carried by the laminate. Hence,

$$\begin{bmatrix} \bar{\sigma}_x \\ \bar{\sigma}_y \\ \bar{\tau}_{xy} \end{bmatrix} = \frac{1}{h} \begin{bmatrix} A_{11} & A_{12} & A_{16} \\ A_{12} & A_{22} & A_{26} \\ A_{16} & A_{26} & A_{66} \end{bmatrix} \begin{bmatrix} \varepsilon_x^0 \\ \varepsilon_y^0 \\ \gamma_{xy}^0 \end{bmatrix} \tag{3.36}$$

This is equivalent to multiplying each term in the (A) matrix by $1/h$ so that, for a symmetric laminate, the average laminate stress values are obtained by multiplying each term of the $(\bar{Q})$ transformed reduced stiffness matrix by V_L, where V_L represents the volume fraction of a particular lamina in the laminate.

Thus the stress strain equations for a symmetric laminate subjected to in-plane loads can be written

$$\begin{bmatrix} \bar{\sigma}_x \\ \bar{\sigma}_y \\ \bar{\tau}_{xy} \end{bmatrix} = \begin{bmatrix} \bar{Q}_{L11} & \bar{Q}_{L12} & \bar{Q}_{L16} \\ \bar{Q}_{L12} & \bar{Q}_{L22} & \bar{Q}_{L26} \\ \bar{Q}_{L16} & \bar{Q}_{L26} & \bar{Q}_{L66} \end{bmatrix} \begin{bmatrix} \varepsilon_x^0 \\ \varepsilon_y^0 \\ \gamma_{xy}^0 \end{bmatrix} \tag{3.37}$$

Where $\bar{Q}_{Lij} = \bar{Q}_{Aij}V_A + \bar{Q}_{Bij}V_B + \bar{Q}_{Cij}V_C$, and V_A, V_B, V_C, etc., represent the relative volume fractions of laminae A, B, C, etc., in the laminate and $\bar{Q}_{Aij}$, $\bar{Q}_{Bij}$, $\bar{Q}_{Cij}$, etc. (which depend on the fibre orientation in each lamina) are given by equations (3.17–3.22). Equation (3.37) relates the average in-plane stresses carried by the laminate to the corresponding mid-plane strains and the laminate can thus be regarded as a homogeneous anisotropic plate. This is a consequence of its symmetric construction, which eliminates

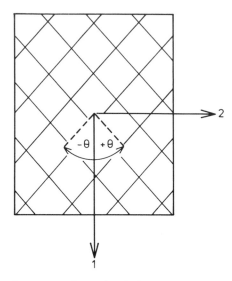

Figure 3.18. Orthotropic symmetric angle-ply laminate.

coupling between in-plane strains and bending moments. Also for a $\pm\theta$ symmetric laminate the Q_{16} and Q_{26} terms sum to zero because they are positive or negative, depending on whether θ is positive or negative, also, the total volume fraction of the $-\theta$ layers is the same as that for the $+\theta$ layers. In the case of $0°/90°$ cross-ply symmetric laminates the $\bar{Q}_{16}$ and $\bar{Q}_{26}$ terms are zero. Hence equation (3.11) can be regarded as the constitutive equation for symmetric laminates and the engineering constants can be obtained from the $\bar{Q}_{Lij}$ terms using equations (3.12).

In Figure 3.13 the elastic properties of a symmetric $\pm\theta$ cross-ply laminate loaded in tension in the 1 direction (see Figure 3.18) are compared with the elastic properties of a single lamina loaded in tension at corresponding off-axis loading values of θ (see Figure 3.12). Considerable increases in v_{xy} and G_{xy} occur in the case of the laminate. The numerical value of E_1 is also maintained to a higher value of θ than is the case during off-axis loading of a single ply. The basic assumption made in developing the constitutive equations for a laminate is that in-plane strains are the same for each lamina. Since their elastic properties are different, the in-plane stresses are different in different laminae and interlaminar shear stresses are developed. By using plane stress conditions to develop the constitutive equations for a laminate the assumption is made that normal stresses between the laminae are zero. This assumption is not justified near a free edge of a laminate where τ_{xz} becomes very large. The high interlaminar stresses tend to cause interlaminar splitting but can be controlled by the stacking sequence of the laminae within the laminate (Pagano and Pipes, 1971), (Curtis 1984).

3.9 ALTERNATIVE COMPOSITE CONSTRUCTIONS CONTAINING PLANAR FIBRE ARRAYS

In §3.8 the properties of laminates consisting of a bonded stack of individual laminae within which the fibres were orientated in various directions were discussed. Other constructions in which the fibres lie effectively in one plane are also used. Some fibre-reinforced composites are constructed from fibres of finite length, usually in the form of fibre bundles. These are distributed randomly in a plane and are impregnated with a polymeric matrix. Their elastic properties can be estimated by assuming that they are effectively constructed from a large number of microlaminae whose orientations are uniformly distributed over all possible angles within the plane of the composite. This assumption ignores the effects associated with the ends of individual fibres or fibre bundles. The effective in-plane elastic modulus is thus obtained by summing the calculated elastic modulus of a microlaminate over all orientation angles.

Fibre bundles are often woven to produce a textile material which is then impregnated with resin. Again the fibres lie primarily in a single plane so that the structure can be regarded as a cross-ply laminate. The degree of curvature of the fibres within the plane of the structure depends upon each particular type of textile material. For example, a higher proportion of straight fibres is present in a satin weave than in a simple woven fabric. Corrections can be applied for the proportion of the fibres that are curved within the plane of the laminate; the problem has been examined by Ishikawa and Chou (1982).

3.10 ELASTIC PROPERTIES OF COMPOSITES REINFORCED WITH DISCONTINUOUS FIBRES

Most fibre composite materials contain fibres of non-uniform strength, so that during the loading of the composite the fibres fracture at the most severe flaws and become discontinuous. Also, since the reinforcing fibres are normally brittle, fibres may be fractured both before and during fabrication. Furthermore, some fibrous composites are constructed for various reasons from fibres having finite and fairly well defined lengths. It is therefore necessary to consider the effects of discontinuous fibres on the mechanical properties of a composite.

In §§3.1 to 3.8 we have assumed that the reinforcing fibres are continuous throughout the whole length of the composite structure so that, when an external tensile load is applied, both fibres and matrix experience equal tensile strains. When the fibres are not continuous, a discontinuity in the strain field must occur near the fibre ends and in these regions the longitudinal strain carried by fibre is less than that carried by the matrix. Hence the average strain carried by a discontinuous fibre is less than the average strain carried by the composite. Discontinuous fibres are thus less effective than continuous fibres in reinforcing a matrix, since over a portion of their length they are carrying a reduced tensile load. The reinforcement "efficiency" of discontinuous fibres falls as their length-to-diameter ratio, or aspect ratio, decreases. The failure of an individual fibre leads to a local redistribution of stress. This may be associated with interfacial bond failure, local plastic deformations or the initiation of a matrix crack.

3.11   STRESS TRANSFER BETWEEN FIBRES AND MATRIX

The tensile stress distribution along a fibre near its end can be obtained approximately using shear lag analysis if the following assumptions are made. The treatment that follows is due to Cox (1952), but similar analyses have been carried out by others. A review of these is given by Hollister and Thomas (1966).

The fibres are assumed to be unidirectionally aligned and dispersed in a regular square or hexagonal array. A load applied to the composite in the fibre direction produces a general strain ε in the composite. One fibre is considered to be discontinuous and each end relaxes elastically, causing local changes in strain to occur in the approximately cylindrical body of matrix material that fills the space between it and its intact nearest neighbours. The diameter of this cylinder is taken as $2r_0$ where r_0 is the distance between adjacent fibre centres. The surface of this cylinder carries the general composite strain ε. Within this cylindrical region the broken fibre and matrix are assumed to behave as elastic materials and to be well bonded together so that there is no discontinuity of strain at the interface. The fibre is assumed to have a much higher elastic modulus than the matrix but the lateral contractions of fibre and matrix are assumed equal. Stress transfer to the broken fibre takes place in a region near the fibre ends as a consequence of the deformations generated in the cylindrical volume of matrix surrounding the ends of the broken fibre (Figure 3.19).

The transfer of load from matrix to fibre at a distance x from the free end of the fibre will depend at that point on the relation between u, the displacement in the matrix in the direction of the fibre axis, and v, the corresponding displacement that would be generated at the same point if the

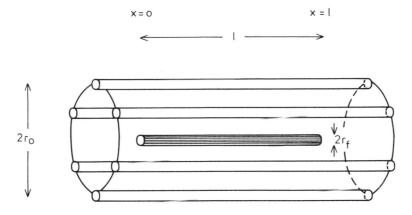

Figure 3.19. Element of composite used to derive the theory of elastic stress transfer.

fibre were removed completely from the cylinder of matrix, i.e. on the displacement of the matrix caused by the presence of the fibre. Hence, if P is the load in the fibre, we have

$$\frac{dP}{dx} = H(u - v) \tag{3.38}$$

where H is a constant depending on the properties of fibres and matrix and their geometrical arrangement.

Differentiation gives

$$\frac{d^2P}{dx^2} = H\left(\frac{du}{dx} - \frac{dv}{dx}\right) = H\left(\frac{P}{E_f A_f} - \varepsilon\right) \tag{3.39}$$

where A_f is the cross-sectional area of the fibre and E_f its Young's modulus. This equation has the solution $P = E_f A_f \varepsilon + R \sinh \beta x + S \cosh \beta x$, where R and S are constants. If the broken fibre has a total length l we have $P = 0$ when $x = 0$, and when $x = l$ and the distribution of tensile stress along the fibre $\sigma_f = (P/A_f)$ is given by

$$\sigma_f = E_f \varepsilon \left(1 - \frac{\cosh \beta(l/2 - x)}{\cosh (\beta l/2)}\right) \tag{3.40}$$

where

$$\beta = \left(\frac{H}{E_f A_f}\right)^{1/2} \tag{3.41}$$

The strain field is symmetrical about the fibre, so that by equating the shear forces τ_r at a distance r from the fibre with those τ_f exerted at the fibre surface, we have per unit length of fibre

$$2\pi r \tau_r = 2\pi r_f \tau_f$$

where $2r_f$ is the fibre diameter, so that $\tau_r = \tau_f r_f / r$. The rate of change of fibre stress dP/dx is set by τ_f, the shear stress transfer at the fibre surface, so that

$$\frac{dP}{dx} = 2\pi r_f \tau_f \qquad (3.42)$$

The shear strain in the matrix at a distance r from the fibre axis can be taken as dw/dr and

$$\frac{dw}{dr} = \frac{\tau_r}{G_m} = \frac{\tau_f r_f}{G_m r} \qquad (3.43)$$

where G_m is the shear modulus of the matrix. The matrix displacement is equal to u at the fibre surface and to v at the position of the adjacent fibres (where $r = r_0$). Hence, after integrating equation (3.43) from r_f to r_0 and substituting from equation (3.38) we have

$$H = 2\pi G_m / \ln (r_0/r_f) \qquad (3.44)$$

so that

$$\beta = \frac{2\pi G_m}{A_f E_f r_f^2 \ln (r_0/r_f)} \qquad (3.45)$$

Since $P = \pi r_f^2 \sigma_f$ we have, from equations (3.40) and (3.42),

$$\tau_f = E_f \varepsilon \left(\frac{G_m}{2E_f \ln (r_0/r_f)} \right)^{1/2} \frac{\sinh \beta(\tfrac{1}{2} l - x)}{\cosh \tfrac{1}{2} \beta l} \qquad (3.46)$$

The average tensile stress $\bar{\sigma}$ carried by a fibre is given by

$$\bar{\sigma} = E_f \varepsilon \left(1 - \frac{\tanh (\tfrac{1}{2} \beta l)}{\tfrac{1}{2} \beta l} \right) \qquad (3.47)$$

so that the contribution of discontinuous fibres to the elastic modulus of a composite approaches that of continuous fibres as $\beta l/2$ becomes large, but tends to zero as $\beta l/2$ becomes very small.

The interfacial shear stress τ_f has a maximum value at the fibre ends and a minimum at the centre of the fibre. When the fibres are long τ_f approaches zero in the central regions of the fibre. Near the fibre ends the analysis predicts very high values for τ_f. The ratio of maximum interfacial shear stress $\tau_{f\,max}$ (which is developed at the end of the fibre) to the maximum fibre tensile stress $\sigma_{f\,max}$ (which is developed at the centre of the fibre), can be obtained from equations (3.46) and (3.40).

For long fibre this is given by

$$\frac{\tau_{f\,max}}{\sigma_{f\,max}} = \left(\frac{G_m}{2E_f \ln (r_0/r_f)}\right)^{1/2}$$

The ratio $\tau_{f\,max}/\sigma_{f\,max}$ is thus dependent on the fibre and matrix properties and the volume fraction of the fibres in the composite. Typically values of the order of 0.1 are predicted for a variety of fibre composite systems. Since $\sigma_{f\,max}$ is large, it follows from this analysis that $\tau_{f\,max}$ will also be large if the strengths of the fibres are to be utilized effectively. If the stress-concentrating effects of the fibre ends are taken into account, even higher values for the ratio $\tau_{f\,max}/\sigma_{f\,max}$ are predicted and this effect is observed under experimental conditions. (See Kelly, 1973)

Rosen (1964) has carried out a similar type of shear lag analysis and obtained an equation of the following form to describe the tensile stress σ_f at any point distant x from the end of the broken fibre as

$$\sigma_f = \sigma_{f\infty} (1 + \sinh \eta x - \cosh \eta x)$$

where

$$\eta^2 = \left(\frac{G_m}{E_f}\right) \left(\frac{V_f^{1/2}}{1 - V_f^{1/2}}\right) \left(\frac{1}{r_f}\right)^2$$

and $\sigma_{f\infty}$ is the stress carried by the fibre at a large distance from the broken end. This analysis predicts stress distribution values similar to those obtained by Cox (1952).

The shear stress values τ_f developed at the fibre matrix interface are now given by,

$$\tau_f = \frac{\sigma_{f\infty}}{2} \left(\frac{G_m}{E_f}\right)^{1/2} \left(\frac{V_f^{1/2}}{1 - V_f^{1/2}}\right) (\cosh \eta x - \sinh \eta x)$$

Thus, when x approaches zero, this analysis also predicts values of τ_f comparable with the maximum tensile stress carried by the fibres.

These analyses are based on the assumption that adhesion is maintained at the interface and that only elastic deformations occur. In practice this is not the case. Since the matrix and the fibre–matrix interface are relatively weak compared with fibre tensile strengths, the large shear stresses developed at the fibre ends can cause shear debonding of the interface, rupture of the matrix or shear yielding of the matrix, depending on the relative stress levels at which these processes occur. Thus the fibre ends become debonded from the matrix and the continuity of elastic displacement, upon which the elastic analysis is based, is lost.

Near the fibre ends the value of τ_f is limited to some characteristic value. If the interface of a polymer matrix composite has failed, stress will continue to be transferred by frictional forces. If the matrix is a metal, yielding can occur very near the fibre surface, so that the shear strength of the interface is now given by the yield stress in shear of the matrix.

For both situations the rate τ of shear stress transfer can be assumed to have a sensibly constant value so that the fibre tensile stress increases linearly with increasing distance from the fibre end. Hence for a cylindrical fibre of diameter $2r_f$,

$$\frac{d\sigma_f}{dx} = \frac{2\tau}{r_f} \tag{3.48}$$

where τ has a numerical value dependent on the particular stress transfer process taking place and which can be determined experimentally. A critical fibre length l_c exists below which stress transfer across the fibre matrix interface is insufficient to cause fibre failure. For a cylindrical fibre this is given by

$$l_c = \frac{r_f \sigma_{fu}}{\tau} \tag{3.49}$$

where σ_{fu} is the ultimate tensile strength of the fibre. (See Kelly, 1973.)

If the fibre is long, stress transfer may take place by elastic displacements in the central regions where the interfacial bond has not been disrupted by high local shear stress values. For most purposes, however, it is sufficient to assume that stress transfer from matrix to fibres takes place only in the vicinity of the fibre ends and occurs at a constant shear stress value. According to this simple model, the maximum distance over which stress transfer takes place at each end of the fibre is $l_c/2$. The situation is illustrated in Figure 3.20. Over this stress transfer distance the average stress carried by the fibre is $\sigma_{fu}/2$ so that the average stress carried by a discontinuous fibre of length l, compared with the same length of a continuous fibre, is given by

$$[\sigma_{fu}(l - l_c) + \sigma_{fu}l_c/2]/\sigma_{fu}l \quad \text{or} \quad (1 - l_c/2l)$$

Since a discontinuous fibre carries a reduced average tensile stress compared with its continuous counterpart, it follows that the elastic modulus of a composite reinforced with aligned discontinuous fibres will be correspondingly reduced. The effect clearly will be small if $l \gg l_c$ and is reduced further at lower fibre stress values when the stress transfer distances are correspondingly reduced. In practice the reduced degree of fibre alignment obtained with discontinuous fibres also has to be taken into account in

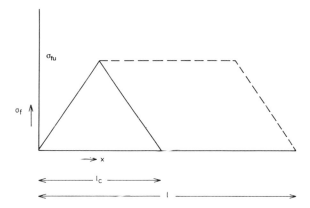

Figure 3.20. Illustrating tensile stress distribution σ_f, developed by a constant value of τ for fibres of different lengths.

considering the contribution of short fibres to the elastic modulus of a composite (Cook, 1968).

Experimental investigations of the reduced values of E_1 obtained for aligned discontinuous carbon fibres in an epoxy resin have been obtained (Dingle, 1974). A reduction in E_1 of 6%, compared with the value obtained for continuous fibres, has been observed (Edwards *et al.*, 1978) when discontinuous fibres having a length of 3 mm were used in a composite. The fibres had diameters of about 8 μm and the estimated fibre critical length l_c was 0.2 mm.

For some applications 3-component composites, consisting of randomly dispersed discontinuous fibres in a polymeric matrix containing a dispersed particulate component, are of interest. Theoretical analyses of such systems have been discussed by Cheng and Weng (1979) and comparisons made with available experimental data.

4
Growth of Matrix Cracks in Brittle Matrix Composites

A high proportion of practical composite materials utilize a matrix that has a lower failing strain than the reinforcing fibres. Polymeric and ceramic matrix composites fall into this category. If a tensile load is applied in the direction of the reinforcing fibres, cracks are generated in the matrix and these are bridged by the reinforcing fibres. If the proportion of reinforcing fibres is small, they may be unable to support the total load carried by the composite material when the matrix fails. Failure of the composite will then occur at the failing strain of the matrix. On the other hand, if the fibres are able to support the total load applied after initial matrix failure, then further matrix cracking will occur. The factors governing this behaviour in unidirectionally reinforced systems are examined in §§4.1 and 4.2.

The failing strain of a brittle matrix is increased by the fibres which, by their presence, modify the rates of release and absorption of energy during the propagation of a matrix crack. This process is discussed in general in § 4.3. and applied to matrix cracks of arbitrary length in §4.4. In §§4.5 and 4.6 these arguments are applied to the growth of cracks in laminates.

4.1 CONDITIONS FOR SINGLE AND MULTIPLE FRACTURE

If the fibres and matrix have the same failing strain, both fracture simultaneously, so that for these conditions the ultimate strength of the composite σ_{cu} is given by

$$\sigma_{cu} = V_f\sigma_{fu} + V_m\sigma_{mu} \qquad (4.1)$$

where V_f and V_m are the volume fractions of fibres and matrix and σ_{fu} and σ_{mu} respectively are their ultimate tensile strengths. In general this condition will not be met. We first consider the matrix to have the lower failing strain and to be elastic up to fracture, so that the stress carried by the composite σ_c as the matrix fails is given by,

$$\sigma_c = V_f E_f\varepsilon_{mu} + V_m\sigma_{mu} \qquad (4.2)$$

where ε_{mu} is the failing strain of the matrix. Here we ignore the mechanics of matrix fracture (discussed in §§4.3 and 4.4), and assume simply that a crack propagates instantaneously through the matrix across the total cross-section of the composite at the critical strain value ε_{mu}. The fibres will fracture when the matrix fails if they cannot support the total load applied to the composite, so that

$$\sigma_c > \sigma_{fu}V_f \qquad (4.3)$$

If the fibres are much stronger than the matrix, this will be observed only at very low values of V_f. For this condition the composite will fracture at a particular cross-section, the process being known as single fracture. As the fibre volume fraction is increased, single fracture will still occur as long as inequality (4.3) holds. At some stage, however, the fibre volume fraction will be sufficient to support all of the load applied to the composite after the matrix has failed, so that

$$\sigma_c < \sigma_{fu}V_f \qquad (4.4)$$

A transverse crack is now formed in the matrix and is bridged by the reinforcing fibres. If we continue to assume that the matrix is not influenced by the presence of the fibres and its failing strain is fairly well defined, transverse matrix cracks will develop elsewhere in the composite structure at negligible increases in the applied stress. Thus, according to this simple model, a transition from multiple to single fracture will take place when the fibre volume fraction becomes sufficiently large to support the total load applied to the composite when the matrix fails. If $E_f > E_m$, the stress carried by the composite at the failing strain of the matrix increases as the fibre volume fraction is increased (Figure 4.1).

A similar argument applies to the alternative condition, in which the Young's modulus of the fibre is less than that of the matrix ($E_f < E_m$). Single fracture of the composite will again occur at the failing strain of the matrix if the fibre volume fraction is so low that the fibres cannot support the additional load applied to them when the matrix fractures. The critical

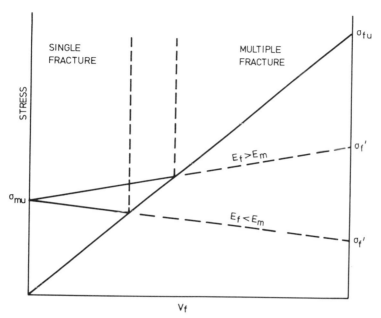

Figure 4.1. Illustrating single and multiple matrix fracture. (Low-failing-strain matrix.) E_f = fibre elastic modulus; E_m = matrix elastic modulus. (After Kelly, 1973.)

fibre volume fraction for the transition to multiple matrix fracture is given for both situations by

$$V_f\sigma_{fu} - V_f\sigma_f' = V_m\sigma_{mu} \qquad (4.5)$$

where σ_f' is the stress carried by the fibres when the matrix fractures. Since $V_m = (1 - V_f)$ we have for the fibre volume fraction at the transition from single to multiple fracture:

$$V_f = \sigma_{mu}/(\sigma_{mu} + \sigma_{fu} - \sigma_f') \qquad (4.6)$$

If $E_f < E_m$, the Young's modulus of the composite will fall as the fibre volume fraction increases so that the matrix now fractures at a diminishing composite stress value as the fibre volume fraction is increased. The matrix is again assumed to fail at the same well defined strain value and σ_{fu} is assumed greater than σ_{mu} even though $E_f < E_m$. The condition is also illustrated in Figure 4.1. Since, when the matrix fractures, the fibre strain is ε_{mu}, the critical fibre volume fraction and strain applied to the composite at the onset of multiple matrix fracture is given by the following relationship for all conditions:

$$V_f E_f \varepsilon_{mu} + (1 - V_f)E_m\varepsilon_{mu} = V_f E_f \varepsilon_{fu} \qquad (4.7)$$

Hence at the onset of multiple matrix fracture,

$$V_f = \left[1 + \frac{E_f}{E_m} \left(\frac{\varepsilon_{fu}}{\varepsilon_{mu}} - 1 \right) \right]^{-1} \tag{4.8}$$

Single fracture of the composite or multiple fracture of the reinforcing fibres can occur in a composite consisting of low failing strain fibres encapsulated in a strong deformable matrix, such as a metal. If the fibres have a much higher ultimate tensile strength than the matrix and their volume fraction is high, we can have $\sigma_{fu} V_f \gg \sigma_{mu} V_m$. For this condition the ultimate failure stress σ_{cu} of the composite is given by

$$\sigma_{cu} = \sigma_{fu} V_f + \sigma'_m (1 - V_f) \tag{4.9}$$

where σ'_m is the stress carried by the matrix at the ultimate tensile failing strain of the fibres. Failure of the fibres now causes single fracture of the composite.

On the other hand, if V_f is small, the load applied to the composite may be supported by the matrix after fracture of the fibres. The matrix will be carrying a stress σ'_m at the failing strain of the fibres so that, if their failure

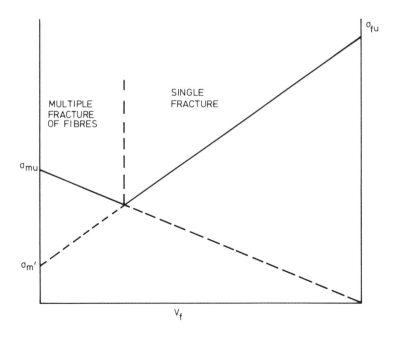

Figure 4.2. Illustrating single and multiple fracture of fibres. (High-failing-strain matrix.) (After Cooper, 1971.)

is not to cause single fracture of the composite, the matrix must be capable of supporting the additional load transferred to it after failure of the fibres. The volume fraction V_{min} of fibres below which this condition will be obtained is given by

$$V_{min} = \frac{\sigma_{mu} - \sigma'_m}{\sigma_{fu} + \sigma_{mu} - \sigma^1_m} \tag{4.10}$$

where σ_{mu} is the ultimate failing stress of the matrix (Kelly and Davies, 1965). This relationship is illustrated in Figure 4.2. For $V_f < V_{min}$ the ultimate tensile strength of the composite is controlled by the strength of the matrix and is given by $\sigma_{cu} = \sigma_{mu} V_m$.

4.2 MULTIPLE FRACTURE OF A BRITTLE MATRIX COMPOSITE

Providing the fibres are able to support the total load applied to the composite, the low-failing-strain matrix of a simple rectangular specimen will eventually develop a series of parallel cracks as the load is increased. The matrix cracks will be bridged by the fibres. Very high shear stresses are developed between the fibres and the matrix where the fibres emerge from the crack face (Dow, 1963; Gray, 1984). The shear stress values attained depend on the difference in elastic modulus between fibres and matrix and on the fibre volume fraction. They are augmented by geometrical stress concentrations located at the point at which the fibres emerge from the face of the matrix crack, being increased as the radius of curvature of this region of transition is decreased. Since the local shear stresses developed are at least of the same order as the stress carried by the fibres, failure of the matrix or the interface occurs at this point and the continuity of elastic displacement at the interface is lost. In the case of brittle matrix composites this results in local debonding of the interface, providing the energy required to destroy the interfacial bond can be supplied by the loading system or by the relaxation of stress in the composite material.

Shear stress transfer between fibres and matrix near to the crack face may be generated after debonding by frictional forces that can be augmented by mechanical interlocking effects due to irregularities in the fibre profile. The numerical value of this shear stress transfer rate is not readily calculable for a practical system but for most purposes can be assumed to be constant. Its effective value τ can be deduced from observations of the experimental behaviour of composite systems.

The separation of the cracks in a brittle matrix having a well defined failing strain is determined by the mechanism of stress transfer between the

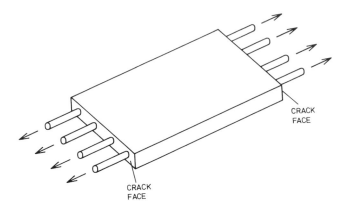

Figure 4.3. Illustrating load transfer from crack-bridging fibres to matrix.

fibres and the matrix. At the position of a crack the reinforcing fibres are carrying the whole load applied to the composite and are thus carrying a maximum tensile strain. At the position of the crack face the matrix is carrying zero stress. In forming the crack the two portions of matrix block on each side of the crack relax elastically from their initial strain value. Hence, differential movement occurs at the fibre–matrix interface. The simplest stress transfer mechanism that can be considered is one in which the shear stress transfer per unit area of fibre–matrix interface is constant. This corresponds to a simple frictional interaction. For this condition the distance of separation of the parallel matrix cracks has been derived by Aveston *et al.* (1971). We consider a rectangular block of matrix traversed by a parallel array of fibres (see Figure 4.3). The stress transferred from one fibre to the matrix over a distance x is $2\pi r \tau x$ where $2r$ is the fibre diameter. If there are N fibres per unit cross-section of the composite the fibre volume fraction, V_f, is given by $N\pi r^2$. Since the load carried by the matrix block at a distance x from the crack must be equal to the load transferred to it from the fibres over this distance, we have

$$2\pi r N \tau x = V_m \sigma_m$$

where σ_m is the tensile stress carried by the matrix at a distance x from the crack face. At some point the load transferred from the fibres to the matrix is sufficient to raise the strain of the latter to that of the fibres. There will be no further stress transfer across the fibre matrix interface beyond this point and at the centre of the block both fibres and matrix carry the same strain (Figure 4.4).

As the load applied to the composite is increased, stress transfer to the

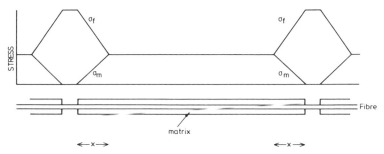

Figure 4.4. Illustrating stress carried by fibres and matrix between adjacent matrix cracks.

matrix takes place over an increasing length of fibre–matrix interface and the load carried by the central region of the matrix block increases. The block will fracture if the fibre stress transfer length is sufficient to allow the block to reach its fracture stress in this region. The length of block at which this will just occur is $2x_1$, where

$$x_1 = \frac{V_m}{V_f} \frac{\sigma_{mu}}{2\tau} r \qquad (4.11)$$

and σ_{mu} is the matrix fracture stress.

Since the value of τ is assumed constant, the stress carried by the fibres decreases linearly from a maximum value at the crack face to a minimum value at the centre of the block of matrix (Figure 4.5). If the block is $2x_1$ wide the fibres will be carrying a strain of ε_{mu}, the failing strain of the matrix, at the centre of the block. Also the fibres will be carrying an additional load $\varepsilon_{mu} E_m V_m$ where they bridge the matrix crack. The additional stress carried by the fibres at this point is therefore $\varepsilon_{mu} E_m V_m / V_f$ and the additional strain is

$$\varepsilon_{mu} E_m V_m / E_f V_f \qquad \text{or} \qquad \varepsilon_{mu}\alpha \quad \text{where} \quad \alpha = E_m V_m / E_f V_f.$$

Hence the average strain carried by the fibres on traversing the block is given by $\varepsilon_{mu}(1 + \alpha/2)$.

If the matrix block has a width slightly greater than $2x_1$ the simple theory predicts that it will fracture, forming two blocks having width of approximately x_1. For this condition the average strain carried by the fibre as it traverses the block will be given from Figure 4.5 by $\varepsilon_{mu}(1 + 3\alpha/4)$. The total extension of the composite is controlled by the elongation of the continuous fibres. Hence the total strain carried by the composite ε_{mc} at the limit of multiple matrix cracking must lie in the range

$$\varepsilon_{mu}(1 + \alpha/2) < \varepsilon_{mc} < \varepsilon_{mu}(1 + 3\alpha/4) \qquad (4.12)$$

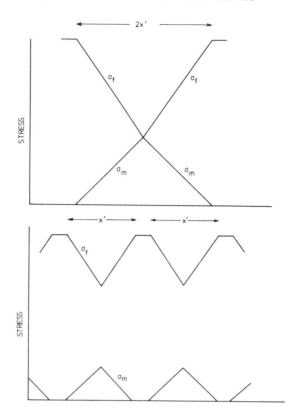

Figure 4.5. Stress distributions at crack spacings of x, and $2x$. (After Aveston *et al.*, 1971.)

If the load is increased further, the reinforcing fibres elongate elastically but the loads transferred to the matrix blocks and the strain distributions within them remain fixed, because the mechanism of load transfer from the fibres to the matrix is not changed. During this extension the elastic modulus of the composite becomes $E_f V_f$. Eventually, the reinforcing fibres reach their ultimate failing strain ε_{fu} at the points where they bridge the matrix cracks. The strain carried by the composite is less than this value because of the reduced strain carried by the fibres within the blocks of matrix.

The composite will eventually fail at a composite stress $\sigma_{fu} V_f$ and a corresponding peak fibre strain of ε_{fu}. Assuming that stress is transferred back to the matrix blocks at the same rate as previously considered, the average strain carried by the fibres at the centre of the block will be $\varepsilon_{fu} - \varepsilon_{mu}\alpha/2$ for blocks $2x_1$ wide and $\varepsilon_{fu} - \varepsilon_{mu}\alpha/4$ for blocks of width x_1. Hence the strain carried by the continuous fibres at the composite failure strain ε_{cu} must lie between these limits, so that

$$(\varepsilon_{fu} - \alpha\varepsilon_{mu}/2) < \varepsilon_{cu} < (\varepsilon_{fu} - \alpha\varepsilon_{mu}/4) \qquad (4.13)$$

During tensile loading the fibres extend elastically by an amount greater than the matrix. Hence there is a differential movement at the interface and energy is absorbed frictionally during multiple matrix cracking. If the specimen is unloaded after matrix cracks have formed, a residual permanent extension of the composite must remain because of the work dissipated irreversibly by frictional losses. These energy losses play a part in increasing the cracking strain of a brittle matrix. The energetics of this fracture process are considered in §4.3.

The analysis of multiple matrix cracking given above for frictionally bonded systems has been developed to deal approximately with fibrous composites in which the fibres remain either fully bonded or partially bonded to the matrix following the development of matrix cracks (Aveston and Kelly, 1973). The fully bonded condition can be evaluated using a shear lag analysis similar to that developed by Cox (1952) (§3.10) to deal with the stresses developed near the end of a broken fibre embedded in an elastic matrix. In the case considered here, the unbroken fibre carries an enhanced stress where it bridges the matrix crack. This falls with increasing distance from the crack face by elastic stress transfer to the matrix. If the stress enhancement at a distance y from the crack face is $\Delta\sigma$ and the stress enhancement at the crack face is $\Delta\sigma_0$, it is shown by Aveston and Kelly (1973) that

$$\Delta\sigma = \Delta\sigma_0 \exp(-\phi^{1/2}y) \qquad (4.14)$$

where

$$\phi = \frac{HE_c}{E_f E_m V_m} \qquad (4.15)$$

For the case of cylindrical fibres the value of H is given by

$$H = \frac{2G_m}{r^2 \ln R/r} \qquad (4.16)$$

Where G_m is the shear modulus of the matrix, $2R$ is the distance between the fibre centres and $2r$ is the diameter of the fibres.

In many composite systems the occurrence of a transverse matrix crack will lead to some debonding of the fibre matrix interface near the crack faces. Stress transfer then takes place by frictional effects over the debonded length of fibre and by elastic effects where the fibre matrix interface has not debonded. The interfacial shear stress transfer rate τ can therefore be regarded as constant when the stress transfer takes place by frictional effects. The value of τ is not constant where the interface has not debonded and for this condition is given by

$$\tau = \tfrac{1}{2}r\Delta\sigma_0\phi^{1/2}\exp\left(-\phi^{1/2}y\right)$$

4.3 INHIBITION OF MATRIX CRACKING BY REINFORCING FIBRES

We consider the energy changes associated with the change from an uncracked condition to one in which a single matrix crack has propagated across the entire cross-section of the material, as computed by Aveston *et al.* (1971). Stress transfer is assumed to take place between the fibres and the matrix over a distance x_1 (defined by equation (4.11)) on each side of the crack. From the argument developed in §4.2, the average increase in the strain carried by the composite over the stress transfer distance x_1 due to the presence of the crack is $\varepsilon_{mu}\alpha/2$, so that the work ΔW, done per unit cross-sectional area of composite by the applied stress $E_c\varepsilon_{mu}$ within a volume x_1 on each side of the crack is given by $\Delta W = E_c\varepsilon_{mu}^2\alpha x_1$. Since $E_c = E_fV_f(1+\alpha)$, we have

$$\Delta W = E_fV_mE_m\varepsilon_{mu}^3r\alpha(1+\alpha)/2\tau \tag{4.17}$$

If we assume a simple frictional interface between the fibres and matrix, no work will be expended in rupturing the interface, but frictional losses will occur as the matrix contracts and the crack-bridging fibres extend. We define this energy loss as U_s per unit area of crack face. The matrix loses strain energy ΔU_m as it contracts and the fibres gain strain energy ΔU_f as they extend. In addition, work will be expended in forming the two fracture faces. This will be $2\gamma_mV_m$, where γ_m is the surface energy of the matrix. From energy balance consideration it is clear that if a parallel-sided matrix crack is to extend across the full width of the composite then

$$2\gamma_mV_m + U_s + \Delta U_f \leqslant \Delta W + \Delta U_m \tag{4.18}$$

Since the matrix is brittle it can be assumed to fracture as a consequence of the presence of flaws of finite size (see Chapter 1, §1.2). A further condition for fracture, therefore, is that the associated matrix failing strain must be reached.

The assumption of a constant interfacial shear stress transfer and cylindrical fibres dictates that the fibre strain ε_f and the matrix strain ε_m will change in a linear manner with increasing distance from the crack face. The strain carried by the matrix is ε_{mu} at a distance x_1 from the crack face. Hence at any position x from the crack face the strain carried by the matrix will be $\varepsilon_{mu}x/x_1$. The strain energy within the block over a distance x_1 from the crack face and per unit cross-sectional area is therefore,

$$\int_0^{x_1} \frac{V_m E_m}{2} \left(\frac{\varepsilon_{mu} x}{x_1}\right)^2 dx = \frac{V_m E_m}{6} \varepsilon_{mu}^2 x_1$$

The initial strain energy present in a length x_1 of matrix before cracking occurs is $V_m E_m \varepsilon_{mu}^2 x_1/2$, so that the strain energy released by a length x_1 of the matrix when the crack has been formed is $V_m E_m \varepsilon_{mu}^2 x_1/3$. It follows that the portion of matrix extending over a distance x_1 on each side of the crack will release an amount of strain energy given by

$$\Delta U_m = E_f E_m V_m \varepsilon_{mu}^3 r \alpha / 3\tau$$

The fibres carry a strain ε_{mu} at a distance x_1 from the crack face and a strain $\varepsilon_{mu} + \alpha \varepsilon_{mu}$ at the crack face. Hence at a distance x from the crack face the fibre strain ε_x is given by

$$\varepsilon_x = \varepsilon_{mu} \alpha (1 - x/x_1) + \varepsilon_{mu}$$

and the strain energy in the fibres over a distance x_1 is given by

$$\frac{1}{2} E_f V_f \int_0^{x_1} \varepsilon_x^2 \, dx = \frac{E_f V_f}{2} \varepsilon_{mu}^2 x_1 \left(\frac{\alpha^2}{3} + \alpha + 1\right)$$

The corresponding initial strain energy carried by the fibres before matrix cracking occurs is $E_f V_f \varepsilon_{mu}^2 x_1/2$, so that the increase in strain energy carried by the fibres over a distance x_1 on each side of the crack is given by

$$\Delta U_f = E_f V_f \varepsilon_{mu}^2 x_1 \alpha (1 + \alpha/3)$$

and

$$\Delta U_f = \frac{E_f E_m V_m}{2\tau} \varepsilon_{mu}^3 \alpha r (1 + \alpha/3)$$

At a distance x_1 from the face of the matrix crack both fibres and matrix are experiencing the same strain ε_{mu} and there is no relative displacement across the interface at this point. The strain carried by the matrix is zero at the crack face, but the fibres here carry a strain $\varepsilon_{mu}(1 + \alpha)$. Hence, at a distance x from the crack face, the difference in the strain carried by the fibres and the matrix is $\varepsilon_{mu}(1 + \alpha)(1 - x/x_1)$. The strain difference decreases linearly to zero as x approaches x_1.

The displacement dm_x of an element of fibre at a position x with respect to the surrounding matrix is obtained by integrating the strain difference from x_1 to x, i.e.

$$dm_x = \varepsilon_{mu}(1 + \alpha) \int_x^{x_1} \left(1 - \frac{x}{x_1}\right) dx = \varepsilon_{mu}(1 + \alpha) \left(\frac{x_1}{2} - x + \frac{x^2}{2x_1}\right)$$

The frictional work done per length dx of fibre is given by $2\pi r \tau \, dm_x \, dx$, so that for the frictional work expended over the length x_1 of fibre we have

$$2\pi r \tau \varepsilon_{mu}(1 + \alpha) \int_0^{x_1} \left(\frac{x_1}{2} - x + \frac{x^2}{2x_1} \right) dx = \pi r \tau \varepsilon_{mu}(1 + \alpha)x_1^2/3$$

The number of fibres per unit cross-section of composite is $V_f/\pi r^2$, so that the total work done per unit cross-section over a distance x_1 on each side of the crack is

$$U_s = 2V_f \tau x_1^2 \varepsilon_{mu}(1 + \alpha)/3r$$

and substituting for x_1 we have

$$U_s = E_f E_m V_m \varepsilon_{mu}^3 \alpha r(1 + \alpha)/6\tau$$

Substituting these values into equation (4.18) we have for the matrix fracture strain ε_{muc}, now modified from its initial value ε_{mu} by the presence of the fibres,

$$\varepsilon_{muc} = \left(\frac{12\tau\gamma_m E_f V_f^2}{E_c E_m^2 r V_m} \right)^{1/3} \tag{4.19}$$

Equation (4.19) indicates that the effective cracking strain of a brittle matrix can be increased above its normal value ε_{mu} if the fibre diameter $2r$ is made small enough; this has been observed experimentally (Cooper and Sillwood, 1972). However, equation (4.19) also indicates that the effective failing strain of the matrix will fall to zero when the fibre volume fraction becomes zero. In reality, the cracking strain of the composite must approach the failing strain of the matrix (controlled by matrix flaws) as the fibre volume fraction tends to zero. Equation (4.19) is not applicable to these conditions because the mechanics of the growth of a matrix crack of finite size is not considered in its derivation.

So far the fibre–matrix interface has been considered to be a simple frictional contact so that no energy is absorbed by debonding prior to the occurrence of frictional losses due to interfacial displacements. Energy must be supplied to rupture a bonded fibre–matrix interface. Let this be W_{id} per unit area of interface. The work done in debonding a length x_1 of a single fibre is therefore $2\pi r x_1 W_{id}$. Since there are N fibres per unit area of fracture surface the work done γ_d in debonding all of the fibres over a distance x_1 on each side of the crack is $4N\pi r x_1 W_{id}$. Substituting for N and x_1 we have

$$\gamma_d = 2V_m \sigma_{mu} W_{id}/\tau$$

Thus, when fibre debonding occurs, it is necessary to add γ_d to the left-hand side of inequality (4.18) so that

$$2V_m(\gamma_m + W_{id}\sigma_{mu}/\tau) < E_c E_f \varepsilon_{mu}^3 \alpha^2 r/6\tau \tag{4.20}$$

The value of W_{id} will vary for different composite systems, being dependent particularly on fibre surface treatment and the presence of interfacial coupling compounds. When a tensile load is applied to a crack-bridging fibre the interface may debond or, depending on the physical parameters of the system, the fibre may fracture before debonding occurs. The mechanics of debonding and fibre extraction have been examined by Bartos (1980) and more recently by Chua and Piggott (1985). The topic has been reviewed by Gray (1984).

4.4 STABILITY OF MATRIX CRACKS OF FINITE LENGTH IN A UNIDIRECTIONALLY REINFORCED BRITTLE MATRIX COMPOSITE

The analysis given in the previous section predicts that the failing strain of a brittle matrix will be increased if it is reinforced with fibres of a sufficiently small diameter. The theory is now extended to take into account the physical processes by which a pre-existing crack of finite size in the matrix will become unstable and propagate transversely to the fibres when the composite is loaded in tension in the direction of fibre alignment (McColl and Morley, 1977a,b). This analysis is based on an approximate estimate of the form of the strain field extending over an appreciable distance on each side of a matrix crack that is bridged by reinforcing fibres. The strain energy released and absorbed during crack extension can be calculated from the form of the strain field. Following Griffith (1920), unstable crack growth is considered to occur when the rate of release of energy with increasing crack length exceeds the rate of energy absorption. The Griffith equation

$$\varepsilon_c = (G/E\pi a)^{1/2} \tag{4.21}$$

gives the critical strain ε_c for the unstable growth of a narrow elliptical crack of length $2a$ in an isotropic elastic sheet under plane stress conditions. The work of fracture of the material is given by G and its Young's modulus by E. At instability the sheet carries a strain ε_c in regions remote from the crack. This strain is generated by a uniform tensile load applied at the edges of the sheet in a direction perpendicular to the length of the crack.

The crack faces carry zero stress and strain energy is released by the elastic relaxation of the material around the crack. Equation (4.21) can be derived from an approximate physical argument if it is assumed that the material has relaxed completely within an elliptical zone around the crack, the size of the zone being equal to twice that of a circle having the crack as a diameter (Figure 4.6). Equation (4.21) is then obtained by equating the rate

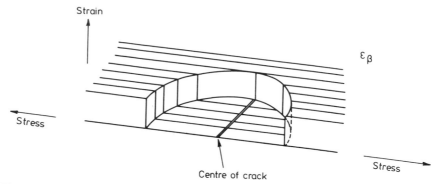

Figure 4.6. Simple physical representation of release of elastic energy around a crack: crack transverse to applied stress; strain measured vertically. One half of the strain field is illustrated. General strain carried by the material given by ε_β.

of release of strain energy with increasing crack length to the rate of absorption of energy through the rupture of the material at the crack tip.

An alternative simple physical model that approximates more nearly to the actual strain field is illustrated in Figure 4.7. This model enables the effect of the addition of crack-bridging fibres to be calculated approximately. The strain is assumed to be zero at the edge of the crack and to increase linearly with increasing distance from the crack face, reaching ε_β on the edge of the elliptically shaped zone surrounding the crack. The strain energy released by this model is computed by dividing the zone into a number of parallel strips each assumed to be independent of each other. Only tensile deformations occurring in the direction of the applied load are

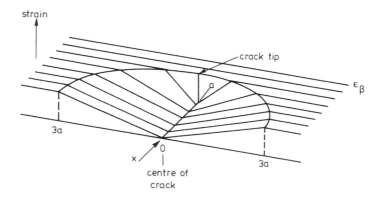

Figure 4.7. Idealized form of strain field around a crack in an unreinforced matrix. (Redrawn from McColl and Morley, 1977a.)

considered in the model. The strain energy contained in one strip is readily calculated and the strain energy contained within the elliptical zone is obtained by summing the strain energy contained within the individual strips. The rate of release of strain energy with increasing crack length is then obtained by numerical differentiation. This is numerically the same as that given by the Griffith analysis if the crack forms the minor axis of an elliptical zone whose major axis is three times the crack length. It will be noted that the strain gradient within the zone increases as the crack tip is approached. The model gives a very simplified version of the strain field and neglects, for example, the localized regions of high stress which would be developed ahead of the crack tips in an elastic solid under these loading conditions.

When crack-bridging fibres are present they carry a maximum strain where they bridge the crack and they also reduce the amount by which the matrix relaxes elastically. The general form of the strain field with fibres present is shown in Figure 4.8.

We now consider the process by which the strain field within the elastic brittle matrix is modified by the presence of crack-bridging fibres. These are assumed to be uniformly distributed, circular in section, and arranged perpendicularly to the crack faces. Stress is transferred between the matrix and the fibres via the fibre–matrix interface at an assumed constant value of shear stress transfer τ. Figure 4.9 shows the strain distribution carried by the reinforcing fibres and the matrix within a section of the composite material aligned perpendicularly to the crack faces and located within the elliptical zone inside which the strain field is perturbed. The slope of the line VQ represents the gradient of strain in the fibres and is given by

$$d\varepsilon_f/dx = -2\tau/E_f r \qquad (4.22)$$

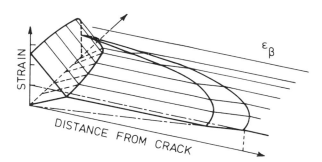

Figure 4.8. Quadrant of strain field around a crack bridged orthogonally by reinforcing members. (Redrawn from McColl and Morley, 1977a.)

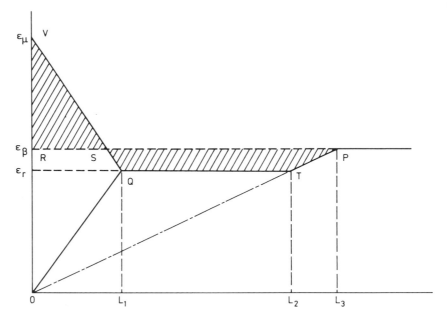

Figure 4.9. Section through quadrant of idealized strain field. (Redrawn from McColl and Morley, 1977a.)

The line OQ in Figure 4.9 represents the gradient of strain in the matrix and is given by

$$d\varepsilon_m/dx = 2V_f\tau/E_mV_mr + \varepsilon_\beta/L_3 \qquad (4.23)$$

L_3 is the distance between any point on the crack to the edge of the elliptical zone measured in a direction parallel to the major axis of the ellipse. The general tensile strain carried by the material outside the elliptical zone, developed as a consequence of the tensile stress applied to the composite structure, is ε_β, so that the strain gradient of the unreinforced matrix is ε_β/L_3.

The model assumes that all changes in strain are confined to the region within the elliptical zone around the crack. It follows that the length of the reinforcing fibres, traversing the elliptical zone and bridging the matrix crack, must have the same overall length as they would have had in the absence of the matrix crack with the composite structure subjected to a uniform tensile strain ε_β. Thus the two areas shown shaded in Figure 4.9 must be equal since these are proportional to the amounts by which different portions of the reinforcing fibre have extended and contracted within the elliptical zone as a consequence of the strain field developed by the matrix crack. From this data values of ε_μ, ε_r, L_1, L_2, L_3, illustrated in Figure 4.9, can be defined. These are

$$\varepsilon_r = \{\varepsilon_\beta L_3/[Q(P + \varepsilon_\beta/L_3)^{-2} + L_3/\varepsilon_\beta]\}^{1/2}$$

$$\varepsilon_\mu = L_1(P + Q + \varepsilon_\beta/L_3)$$

$$L_1 = \varepsilon_r(P + \varepsilon_\beta/L_3)^{-1} \tag{4.24}$$

$$L_2 = L_3\varepsilon_r/\varepsilon_\beta$$

$$L_3 = 3(a^2 - y^2)^{1/2}$$

where $P = 2V_f\tau/E_m V_m r$ and $Q = 2\tau/E_f r$. The distance from the centre of the crack to the segment considered is represented by y and a is the half crack length. The half crack opening at a distance y from the centre of the crack is obtained from the difference in integrated strain between the reinforcing fibres and the matrix over the distance OL_1 and is given by $\mu = \varepsilon_\mu L_1/2$.

These equations enable the strain field within the elliptical zone to be defined. In Figure 4.10 the predictions of the theory are compared with the measured strain field developed around a crack in an aluminium sheet reinforced by steel reinforcing members. The experimental data agree quite well with the strain field predicted by the analytical model. According to the model the strain carried by the fibres bridging the crack can be appreciably

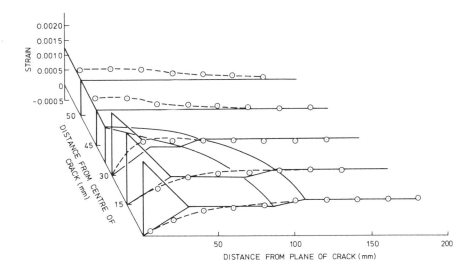

Figure 4.10. Experimental observations (○) of the strain distribution around a crack in an elastic sheet bridged by reinforcing members compared with calculated values. (Redrawn from Morley and McColl, 1984.)

greater than the strain carried by the fibres elsewhere in the system. Hence, a matrix flaw can initiate fibre failure even though the general tensile strain carried by the composite is less than the fibre tensile fracture strain.

From equations (4.24) the strain energy ΔWR_y released by a parallel-sided segment of composite material width Δy and situated within the elliptical zone at a distance y from the centre of a crack can be determined, using procedures similar to those described in §4.3. Following McColl and Morley (1977b), we have

$$\Delta WR_y = [E_c\varepsilon_\beta^2 L_3/2 - E_c\varepsilon_\beta^2(L_3^3 - L_2^3)/6L_3^2$$
$$- E_c\varepsilon_r^2(L_2 - L_1)/2 - V_m E_m\varepsilon_r^2 L_1/6$$
$$- V_f E_f(\varepsilon_\mu^2 + \varepsilon_\mu\varepsilon_r + \varepsilon_r^2)L_1/6]\Delta y \qquad (4.25)$$

The total strain energy released as a consequence of the presence of a matrix crack can be computed by numerically integrating equation (4.25) over the whole of the elliptical zone. This can be done by dividing a quadrant of the ellipse into a number of zones of equal width which are then summed to give the energy released over the entire quadrant. The rate of release of strain energy for increasing crack lengths can then be obtained by numerical differentiation.

We assume a frictional bond between the reinforcing fibres and the matrix, so that displacements occur at the interface at a constant frictional shear stress value τ, and the frictional energy lost is calculated as in §4.3. Frictional losses now occur over a distance L_1 along the fibre, measured from the edge of the crack, and the strain carried by the fibres where they bridge the crack is ε_μ. Substituting these values into the previous computation gives for the frictional work expended over the surface of one fibre: $\pi r \tau \varepsilon_\mu L_1^2/3$.

Since the number of fibres in a section of width Δy and unit thickness is $V_f \Delta y/\pi r^2$ the work done against frictional forces in this segment of the quadrant is $\Delta WA_y = V_f\tau\varepsilon_\mu L_1^2\Delta y/3r$. Substituting for ε_μ from equations (4.24) gives

$$\Delta WA_y = V_f\tau\Delta y(P + Q + \varepsilon_\beta/L_3)L_1^3/3r \qquad (4.26)$$

The total energy absorbed frictionally is obtained by numerically integrating equation (4.26) over the elliptical zone around the crack. This can be computed for incremental increases in crack length, the rate of absorption of energy with increasing crack length then being obtained by numerical differentiation. This is found to increase at a diminishing rate with increasing crack length, eventually reaching an upper limit set by the characteristics of the particular composite system. Energy is also absorbed in rupturing the matrix to form a transverse matrix crack and also in any

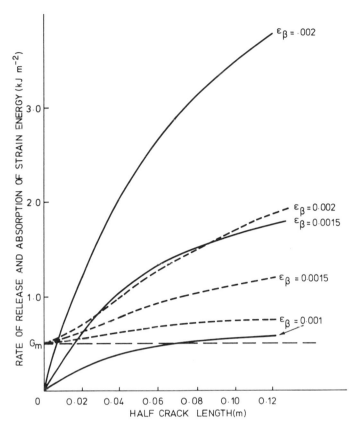

Figure 4.11. Rates of release (full line) and absorption of energy (broken line) with increasing crack length. Matrix work of fracture denoted by G_m. (Redrawn from McColl and Morley, 1977b.)

chemical debonding of the fibre–matrix interface. The rate of release of strain energy with increasing crack length and the rate of absorption of strain energy with increasing crack length can thus be calculated for a particular composite system. A matrix crack of arbitrary length will be on the point of instability if, for a small increase in crack length, the amount of strain energy released equals the amount of strain energy absorbed.

In Figure 4.11 the calculated rates of release and absorption of strain energy with increasing crack length are shown for a polymeric matrix composite reinforced with steel wires. The work of fracture of the matrix has been enhanced by the inclusion of a small quantity of glass fibres. Critical half crack lengths of approximately 0.01 m and 0.02 m are computed for this system for composite tensile strains of 0.002 and 0.0015 respectively. Note that, according to this analysis, there is no critical crack length for this system at a composite strain of 0.001.

In the analysis it is assumed that the elliptical zone around the crack is divided into a finite number of segments, each perpendicular to the length of the crack. Adequate computing accuracy is obtained if the zone is divided into five or more elements. The point is of some significance because flaws of the size expected to be present in a brittle matrix composite (computed on the basis of Griffith crack theory for the unreinforced matrix) would be bridged by several fibres of carbon or glass, which are typically of the order of 10 μm in diameter. It will be noted that if very large reinforcing fibres are considered, e.g. boron, the intrinsic matrix flaw sizes will be of the same order as the fibre diameters. Hence, for these conditions, the strain field theory outlined here would not be expected to give accurate predictions of enhancements in matrix failing strains which, in any event, would be small.

In Figure 4.12 the effect of various fibre volume fractions on the matrix failing strain is shown for three fibre diameters (1 μm, 8 μm and 200 μm). The matrix crack length is assumed to be ~40 μm. The physical validity of the model is again doubtful at extremely small fibre volume fractions, even

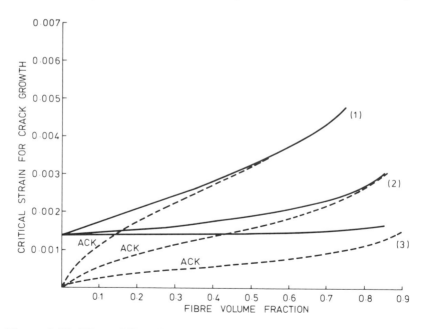

Figure 4.12. Effect of fibre diameter and volume fraction on the critical strain for crack extension. Curve (1) fibre diameter = 1 μm; curve (2) fibre diameter = 8 μm; curve (3) fibre diameter = 200 μm. Predictions of equation (4.19) indicated by curves labelled ACK (Redrawn from Korczynskyj *et al.* 1981.)

for the small-diameter fibres, because of the small number of fibres that would be bridging a crack of the size considered. However, for these conditions the predicted fracture strain enhancement of the matrix by the fibres is again negligibly small. The stability of the matrix crack increases with increasing fibre volume fraction for all fibre thicknesses but is much more pronounced for the thinner fibres. In Figure 4.12 a comparison is made with the theory developed in §4.3. It will be noted that for the thickest fibres considered here (200 μm in diameter) the matrix cracking strain predicted by equation (4.19) is less than the intrinsic matrix failing strain at all fibre volume fractions. At high fibre volume fractions and small fibre diameters the predictions of both theories tend to converge. The analysis outlined in this section indicates a continuous relationship between the fibre volume fraction and the matrix failing strain that covers the whole range of fibre volume fractions and predicts, as does equation (4.19), a very considerable enhancement of the matrix failing strain when fibres of small diameter are present in an appreciable volume fraction.

When unstable extension of a matrix crack occurs, the composite will fail by single fracture if the fibres are incapable of supporting the total load then applied to the composite, as described in inequality (4.3). It is apparent from Figure 4.8 that the fibres bridging the centre of a stable matrix crack will be carrying a higher strain than fibres elsewhere in the composite material. The consequent fracture of a sufficient number of crack-bridging fibres may precipitate failure of the composite by destabilizing the matrix crack and causing sequential failure of the remaining crack-bridging fibres. Theoretically this mechanism can cause the composite to fail at a lower strain than that predicted by inequality (4.3); this mechanism is discussed at greater length in §5.4.3.

4.5 MULTIPLE CRACKING IN CROSS-PLY LAMINATES

Because of the severe anisotropy in the elastic properties of unidirectionally aligned fibrous composites, these materials are often used in the form of laminates. The elastic characteristics of the laminate are then controlled by the orientation of the fibres in the various laminae, as discussed in §3.8. The strength of a lamina is low when a tensile load is applied perpendicular to the direction of the fibre alignment, and cracks propagate parallel to the fibres at low transverse strains. In the case of simple cross-ply laminates, cracking occurs in the transverse plies at strains very much less than the failing strain of the longitudinal plies. Also the strength of the ply in the longitudial direction is much greater than its strength in the transverse direction. Hence, when the proportion of transverse and longitudinal plies is similar, it follows that the longitudinal plies can continue to support the total load applied to the laminate after the transverse plies have fractured. Under these conditions multiple cracking occurs in the transverse plies. The situation is similar to that of multiple matrix cracking in a fibre-reinforced system as described in §4.2 but with the stress transfer taking place across the inter-ply interface instead of the fibre–matrix interface. The inter-ply interface may or may not become debonded (Figure 4.13(a) and (b)). When debonding occurs, stress transfer between the plies takes place through residual frictional interaction. When debonding does not occur, stress transfer across the inter-ply interface takes place by elastic displacements. The crack is confined to the transverse ply (Figure 4.13(a)). This process has been analysed by Garrett and Bailey (1977). Near the crack the transverse ply deforms in shear in the direction of loading. The displacement increases towards the mid-plane of the ply. This deformation results in the transfer of stress across the inter-ply interface and the magnitude of this stress transfer can be estimated using a shear lag analysis similar to that described in §3.11. At the position of the crack in the transverse ply the longitudinal ply carries the load previously supported by the transverse ply. This additional load falls as the distance y from the crack increases owing to stress transfer across the inter-ply interface. For the three-ply laminate illustrated in Figure 4.14 the additional stress $\Delta\sigma$ carried by the longitudinal ply at a distance y from the crack face is given by Garratt and Bailey (1977) as

$$\Delta\sigma = \Delta\sigma_0 \exp\left(-\phi^{1/2}y\right) \tag{4.27}$$

where

$$\phi = \frac{E_c G_t}{E_1 E_t}\left(\frac{b+d}{bd^2}\right)$$

E_c is the initial laminate Young's modulus in the y direction and can be determined for a cross-ply laminate with reasonable accuracy using a simple rule of mixtures approach. G_t is the transverse shear modulus of the transverse ply and E_1 and E_t the Young's modulus of the longitudinal and transverse plies, respectively, measured in the y direction. The shear stress transfer across the inter-ply interface is given by

$$\tau_i = -b\left(\frac{d\Delta\sigma}{dy}\right) = b\Delta\sigma_0\phi^{1/2}\exp(-\phi^{1/2}y) \tag{4.28}$$

and the load F transferred back to the transverse ply at a distance y from the crack face is given by

$$F = 2bc\Delta\sigma_0[1 - \exp(-\phi^{1/2}y)] \tag{4.29}$$

where c is the width of the laminate (Figure 4.14).

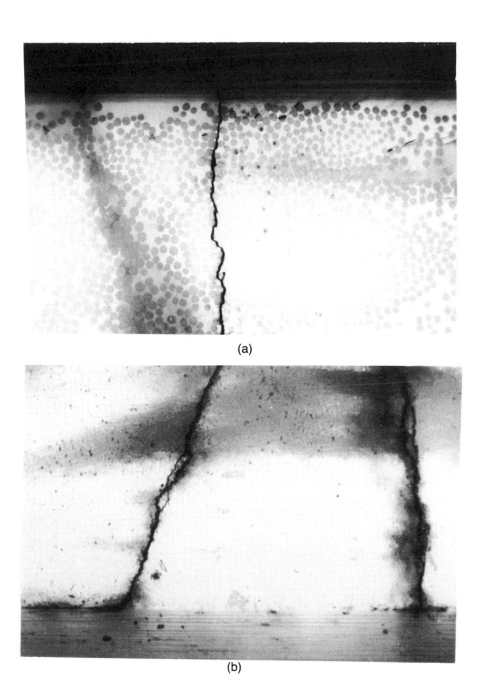

Figure 4.13. Interaction of a transverse crack with a longitudinal ply. There is no evidence of debonding between plies for a transverse ply thickness of 1–2 mm (*a*). Some debonding between plies and 45° oblique cracks are observable for a transverse ply thickness of 4 mm (*b*). (From Parvizi and Bailey, 1978—courtesy of Chapman and Hall Limited.)

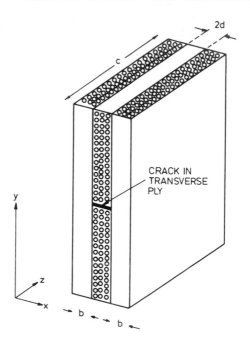

Figure 4.14. Illustrating analysis of cracking in cross-ply laminates. (Redrawn from Garrett and Bailey, 1977.)

The cracking of the transverse plies is inhibited by the presence of the longitudinal plies. The effect has been estimated quantitatively by Bailey *et al.* (1979), using an argument similar to that developed in §4.3 to deal with constrained cracking of a brittle matrix. They calculated the cracking strain ε of the transverse ply in the laminate shown in Figure 4.14 to be given by

$$\varepsilon = [GbE_1 \, \phi^{1/2}/E_t E_c (b + d)]^{1/2} \qquad (4.30)$$

where G is the work of fracture for cracks propagating parallel with the fibres in the transverse ply. A corresponding relationship can be deduced for laminates having a 90°/0°/90° geometry. After curing of the polymeric matrix, differential thermal contraction of the plies occurs, so that the transverse plies of a laminate will generally be supporting an initial thermal tensile strain. Thus the sum of the thermal strain plus the cracking strain generated by an applied load is predicted by equation (4.30). Depending on the geometry of the laminate, the cracking strain of the transverse ply predicted by equation (4.30) can be greater or less than its normal value. The equation is regarded as applicable only when enhanced cracking strain values are predicted.

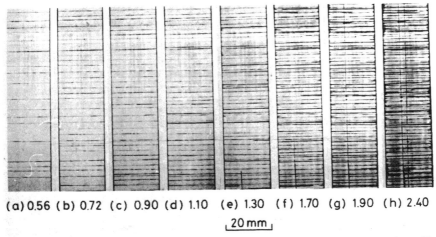

(a) 0.56 (b) 0.72 (c) 0.90 (d) 1.10 (e) 1.30 (f) 1.70 (g) 1.90 (h) 2.40

⌊20 mm⌋

Figure 4.15. Illustrating progressive developments of transverse cracks with increasing strain. (From Manders *et al.*, 1983—courtesy of Chapman and Hall Limited.)

In the case of engineering laminates, fracture of the transverse plies occurs because of the presence of flaws that vary in severity and hence cause failure at different strain values. The first crack to appear in the transverse ply is caused by the most severe flaw present. After a crack has been formed the load carried by the transverse ply is reduced near to the newly formed crack, as described by equation (4.29). More cracks will be formed as the load carried by the laminate is increased, and their location will depend on the distribution of flaws and the distribution of stress in the transverse ply.

Manders *et al.* (1983), point out that the observed distributions of cracks in the transverse ply of a cross-ply laminate are consistent with a statistical variation in the strength of the transverse ply. Figure 4.15 illustrates the observed increase in the number of transverse cracks with increasing composite strain. The spacing of the cracks reaches a limiting value with increasing strain and this value depends upon the geometry of the laminate, the limiting crack spacing increasing as the thickness of the transverse ply increases. This is an expected consequence of a stress transfer mechanism operating at the interface between plies. (Parvizi and Bailey, 1978.)

So far consideration has been given to the development of cracks in the transverse plies propagating in a direction perpendicular to the direction of the applied load. Splitting in the longitudinal plies can also occur parallel to the direction of the fibres and to the applied load. This is seen particularly in glass-fibre-reinforced cross-ply laminates and is due to stresses set up as a consequence of differences in the Poisson ratios of the 0° and 90° plies (§3.8). Thus two sets of cracks at right angles to each other can be produced by an external load applied parallel to the 0° fibres (Bailey *et al.*, 1979; Aveston and Kelly, 1980).

In the case of some laminates, particularly carbon-fibre-reinforced systems, the thermal stresses produced after curing are sufficient by themselves to cause cracking parallel to the fibres as the laminate is cooled to room temperature.

4.6 CRACK INITIATION IN LAMINATES

The analysis of the mechanics of crack extension in a fibre-reinforced low-failing-strain matrix that was given in §4.4 can be modified to deal with multi-ply laminates. The simplest system to consider is the 0°/90° cross-ply system. If a tensile load is applied in the 0° direction the growth of cracks parallel to the fibres in the 90° plies will be inhibited by the presence of the 0° plies. Stress transfer takes place across the interlaminar interface. We consider the growth of a flaw of arbitrary length extending through the full thickness of the 90° low-failing-strain lamina and propagating in a direction parallel with the fibres in this lamina and perpendicular to the interlaminar interface (Figure 4.16).

If the shear stress transfer across the interlaminar interface is assumed to take place at a constant value τ, equations (4.24) dealing with cylindrical fibres in a matrix can be transposed to deal with a laminate using the following substitutions (Korczynsky and Morley, 1981):

$$P \text{ (fibre system)} = \frac{2V_f\tau}{E_m V_m r}$$

$$P \text{ (laminate system)} = \frac{K\tau}{E_t T_t T_{tot}}$$

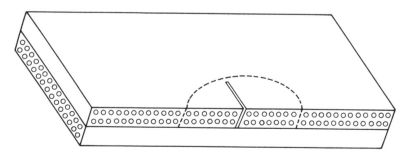

Figure 4.16. Illustrating transverse crack growth in an idealized two-ply 0°/90° laminate. (After Morley, 1983.)

$$Q \text{ (fibre system)} = \frac{2\tau}{E_f r}$$

$$Q \text{ (laminate system)} = \frac{K\tau}{E_1 T_1 T_{\text{tot}}}$$

$$T_1 = \frac{t_1}{t_1 + t_t}$$

where t_1 and t_t are the thicknesses of the longitudinal and transverse plies,

$$T_t = \frac{t_t}{t_1 + t_t}$$

and E_1 and E_t are the Young's modulus values of the individual plies measured respectively parallel to and perpendicular to the fibres in the ply. Here both transverse and longitudinal plies are assumed to have the same mechanical properties. Also, $K = 1$ for a 2-ply laminate; $K = 2$ for a 3-ply laminate; $K = 2M$ for a laminate containing a large number of individual plies; N is the number of longitudinal ($0°$) plies; M the number of transverse ($90°$) plies. $N = M \pm 1$ and $T_{\text{tot}} = NT_1 + Mt_t$.

To transpose from equations (4.24), (4.25) and (4.26), dealing with cylindrical fibres in a matrix, to the mechanics of cracking of the transverse plies of a laminate, we also substitute

$$T_1 \text{ for } V_f, \qquad T_t \text{ for } V_m, \qquad E_1 \text{ for } E_f, \qquad E_t \text{ for } E_m, \qquad K/T_1 T_{\text{tot}} \text{ for } 2/r$$

In computing the rate of release of strain energy with increasing crack lengths in the transverse plies (by substitution into equation (4.25)), the elastic modulus of the laminate in the direction of the applied load, E_c, is taken as $(E_1 T_1 + E_t T_t)$. It is assumed that a crack of arbitrary length in a transverse ply will become unstable when the rate of release of strain energy with increasing crack length is equal to the rate of absorption of energy. This can be computed as described in §4.4. In the simple 2-dimensional theory of the mechanics of transverse ply cracking, set out above, it is assumed that crack growth takes place simultaneously in the same plane in all of the transverse plies. In practice cracks of various lengths will be present in various positions in the transverse plies. Limits for unstable crack growth can be obtained readily for simple situations. For example, in the case of a three-ply $0°/90°$ system two arrangements are possible—one in which a single transverse ply is sandwiched between two longitudinal plies ($0°/90°/0°$ system), and one in which a single longitudinal ply is sandwiched between two outer transverse plies ($90°/0°/90°$ system). In Figure 4.17 the computed cracking strains for the central transverse ply in the first

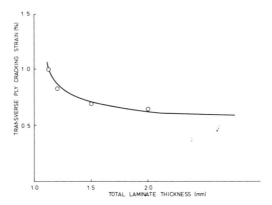

Figure 4.17. Computed transverse ply cracking strains compared with experimental values for 0°/90°/0° system. (Redrawn from Korczynskyj and Morley, 1981.)

arrangement are compared with cracking strain values observed by Bailey *et al.* (1979). The thickness of the outer longitudinal plies is constant but the thickness of the inner transverse ply is varied. The computed cracking strain values are based on the known elastic properties of the laminae and their dimensions. The initial flaw size in the transverse ply is computed from its observed failing strain value, and the computed cracking strain values approach this as the thickness of the central transverse ply (and hence the total thickness of the laminate T_{tot}) is increased. The effective rate of shear stress transfer at the inter-ply interface is assumed constant at $10 \ \mathrm{MN \ m^{-2}}$. Although τ would not be expected to be constant, see equation

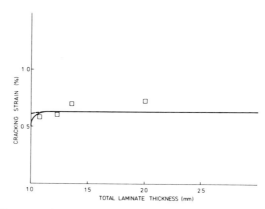

Figure 4.18. Computed transverse ply cracking strains compared with experimental values (90°/0°/90°) system. (Redrawn from Morley, 1983.)

(4.28), this is a reasonable approximation for this analysis since stress transfer takes place only in a localized region near to the crack face (see Figure 4.9).

In Figure 4.18 computed cracking strain values for the second arrangement are compared with the observed values. Here the thickness of the outer transverse plies is held constant as the thickness of the inner longitudinal ply is varied. For this arrangement cracking may occur in one, or both, outer transverse plies, and two computed curves are shown corresponding to these conditions. The two curves are indistinguishable except when the thickness of the central longitudinal ply is very small (Morley, 1983). In both Figures 4.17 and 4.18 the same parameters are used in the computation of the theoretical cracking strain values.

5

Composite Fracture under Tensile Loading

In this chapter we consider the processes occurring as a fibre-reinforced material becomes separated into two or more fragments under the action of an external tensile load. The necessary work of fracture is supplied either by the displacement of the point of application of the external load or by the elastic relaxation of the material as a crack, or region of damage, extends. The work of fracture G_{1c} of a fibrous composite for cracks propagating in a crack opening mode in a direction perpendicular to the alignment of the fibres can be very high. It cannot be accounted for simply by adding together the contributions of the fibres and matrix. For example, polymeric materials have G_{1c} values of the order of $100 \, \mathrm{J \, m^{-2}}$ and brittle fibres of the order of $10 \, \mathrm{J \, m^{-2}}$, whereas the value for the composite may be of the order of $10^5 \, \mathrm{J \, m^{-2}}$. The high work of fracture of the composite is due to interactions between fibres and matrix that lead to the absorption of large amounts of energy. These energy-absorbing processes are controlled by the characteristics of the fibre–matrix interface.

The work of fracture can be readily measured during controlled crack propagation and equated with the nominal surface energy γ of the crack faces or with the work G_{1c} done in propagating the crack through a unit cross-section of the composite such that $G_{1c} = 2\gamma$. Controlled fracture is usually achieved using notched specimens loaded in bending or alternatively with double cantilever specimens (Figure 5.1). In both cases the crack extends at a diminishing load level, although the displacement of the points of application of the load increases with crack extension. In this way

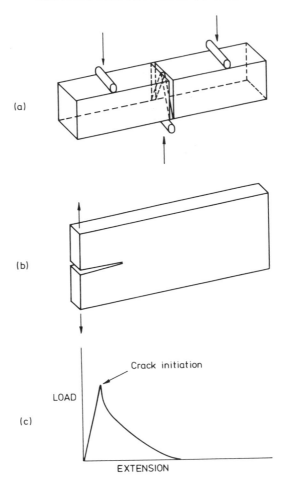

Figure 5.1. (*a*) Notched specimen loaded in bending. (After Tattersall and Tappin, 1966.) (*b*) Double-cantilever specimen. (*c*) Typical relationship between the applied load and its displacement during crack growth.

controlled stable crack growth can be achieved even in the case of very brittle materials. Tensile crack opening forces are developed in both types of specimen. Since the strain energy contained within the loading system is zero both at the start and at the end of the experiment, the work done on the specimen is given simply by the area under the load–displacement curve of the testing machine. This can be related to the work of fracture of the material from a knowledge of the cross-sectional area of the fractured specimen.

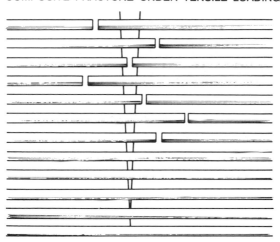

Figure 5.2. Schematic view of the region near the tip of a crack in a brittle matrix composite.

In the case of brittle matrix materials, in which the failing strain of the fibres may be higher than that of the matrix, it is observed that fibres remain intact, bridging the matrix crack. These fibres may fracture at some distance from the crack plane and may then be pulled out of the matrix as the crack faces separate. Thus the region of damage may extend over an appreciable volume of the composite structure (Figure 5.2). Conversely, where the matrix has a much higher failing strain than the fibres, as is the case if the matrix is a ductile metal, the fibres may fracture sequentially more or less in a single plane leaving behind bridges of matrix material that in turn deform and fracture as the crack propagates. A number of energy-absorbing mechanisms have been identified that operate as a consequence of a matrix crack being bridged by intact fibres. This is the normal model of failure of fibre-reinforced polymers and ceramics (see Figure 5.2). As the crack faces separate, the load carried by the crack-bridging fibres increases and the interface between the fibres and the matrix debonds with consequent absorption of energy. Energy can also be absorbed by frictional interactions occurring at the debonded interface due to displacements between matrix and fibres as the load on the fibres is increased. Since the crack-bridging fibres carry a higher stress than the rest of the material, they absorb strain energy. If they are discontinuous, or fracture sufficiently near the crack plane, they can be extracted from the matrix against interfacial frictional effects and this provides a further energy-absorbing mechanism. Alternatively, the highly stressed crack-bridging fibres may fracture at the crack plane and relax elastically back into the matrix. The strain energy they held at fracture is then dissipated within the composite structure and is not available to assist the propagation of the primary crack. Broken or discontinuous fibres may also be extracted from a ductile matrix. If they are strongly bonded to the matrix, work is expended in deforming the matrix in shear near to the fibre surface as the fibres are extracted.

In this chapter we deal firstly with the various energy-absorbing processes that occur as a consequence of interactions between fibres and matrix during crack extension. The general principles of these processes are now well understood. We then deal with the initiation of composite failure considered as being due to the cumulative failure of the fibres at random positions within the composite structure with increasing values of composite strain. This is caused by a distribution of flaws of differing severity along the lengths of individual fibres. Various processes have been proposed by which composite failure is initiated by fibre failure. Early theories assumed that failure would occur when insufficient intact fibres remained within some particular cross-section to support the applied load. More recently the enhancement of stress carried by the intact fibres due to the fracture of neighbouring fibres has been taken into account. Finally, in this chapter, some consideration is given to the initiation of fracture in hybrid fibre composite systems. Except where specified, the following discussions relate to unidirectionally reinforced systems loaded in tension in the direction of fibre alignment.

5.1 THE STRENGTH OF COMPOSITES REINFORCED WITH DISCONTINUOUS FIBRES

We assume that localized enhanced stresses are not generated in fibres that are adjacent to the end of a discontinuous fibre, so that the strength of a composite, containing unidirectional discontinuous fibres loaded in the direction of the fibre alignment, is given by

$$\sigma_c = \bar{\sigma}_f V_f + \sigma_m^1 V_m \tag{5.1}$$

where the mean stress carried by the fibres at composite failure is $\bar{\sigma}_f$ and σ_m^1 is the stress carried by the matrix at the failure strain of the composite. The average stress carried by the fibres depends upon their aspect ratio (length-to-diameter). If we assume that to a first approximation the stress transfer across the fibre matrix interface takes place at a constant value τ then for the fibre critical length l_c we have (see §3.11)

$$\sigma_{fu} = \frac{2\tau l_c}{d} \tag{5.2}$$

where σ_{fu} is the ultimate tensile strength of the fibre and d is its diameter. Thus a discontinuous fibre can be fractured if its length is greater than l_c. Since the stress increases linearly from zero at each end of the fibre to σ_{fu} at the centre, a fibre having a length equal to l_c supports an average stress of $\sigma_{\text{fu}}/2$. If the fibre length l is less than l_c the average stress carried by the fibre will be $\sigma_{\text{fu}}l/2l_c$. The simple linear stress distribution along the fibre considered here could be developed as a consequence of plastic flow in the matrix (when τ will represent the yield strength in shear of the matrix) or by frictional stress transfer at the interface if the matrix is, for example, a brittle polymer (when τ represents the interfacial frictional interaction). Once this linear stress distribution is fully developed, an increase in the tensile deformation of the composite will not cause the stress carried by the fibres to increase. Eventually the matrix will fail at a stress σ_{mu}. Thus the strength of a composite containing such fibres will be given by

$$\sigma_c = \sigma_{\text{fu}} \frac{l}{2l_c} V_f + \sigma_{\text{mu}} V_m \tag{5.3}$$

If the fibres are longer than l_c then the average fibre stress is given by

$$\bar{\sigma}_f = \sigma_{\text{fu}}[1 - (l_c/2l)] \tag{5.4}$$

and the strength of the composites becomes

$$\sigma_c = \sigma_{\text{fu}} V_f [1 - (l_c/2l)] + \sigma_m^1 V_m \tag{5.5}$$

5.2 DUCTILE-MATRIX COMPOSITES—WORK OF FRACTURE

Copper reinforced with strong, brittle, continuous tungsten wires is a typical ductile matrix composite and has been studied in detail by Cooper and Kelly (1967). A region of damage propagates by sequential fibre failure. The extent to which the fracture of one fibre influences the point of failure of the adjacent fibres is controlled by the fibre–fibre coupling and is dependent on the proximity of the fibres and the shear strength of the fibre–matrix interface. When the fibre–matrix interface is strong, an almost planar crack propagates through the array of tungsten wires, the direction of the crack front following the points of minimum separation of neighbouring fibres.

After fibre failure the still-intact ductile matrix bridges carry a stress equal to the yield stress of the matrix. This excess stress is transferred back into the fibres over a certain distance on each side of the plane of fibre fracture. We take the interfacial shear stress τ to be equal to the yield strength in shear of the matrix. The volume of matrix taking part in the deformation process is readily calculated if it is assumed that the stress carried by the matrix outside the stress transfer distance is negligible. This is a reasonable approximation since the stress carried by the matrix will generally be small by comparison with the stress carried by the fibres at their failing strain. Thus we assume that the load carried by the matrix bridges at the yield strength of the matrix, $(V_m\sigma_{mu})$ is transferred to the fibres over a distance x on each side of the fracture plane by a shear stress τ acting at the fibre–matrix interface, so that

$$V_m\sigma_{mu} = N\pi d\tau x$$

where σ_{mu} is the tensile yield strength of the matrix, N is the number of fibres per unit cross-section of the composite, and d is the diameter of the fibres. Hence

$$x = \frac{V_m\sigma_{mu}}{4V_f\tau} \tag{5.6}$$

The volume of matrix material in the bridges on one side of the fracture face is given by xV_m per unit area of fracture surface. The work done in breaking the bridges can be assumed to a first approximation to be equal to that required to deform a similar volume of matrix material to its ultimate tensile strength.

If this is W_m then the effective surface energy of one crack face due to the plastic deformation of the matrix bridges γ_{mp} is given by

$$\gamma_{mp} = \frac{V_m^2\sigma_{mu}dW_m}{4V_f\tau} \tag{5.7}$$

and the composite work of fracture due to the deformation of the matrix bridges is $2\gamma_{mp} = G_{mp}$. The contribution of the plastic matrix to the work of fracture of the composite is thus proportional to the fibre diameter.

The work of fracture of the fibres themselves has also to be included in the work of fracture of the composite. If the fibres are brittle, the surface energy of the fractured ends of the fibres will be negligible. If the fibres are ductile wires, they deform plastically before fracture and, because of the constraint of the matrix, the yield strength of the wire may be enhanced. It is observed that the length of wire that deforms at the crack face is only of the order of the wire diameter but, since the ultimate tensile strength of the wire can be large, the contribution of ductile wires to the work of fibre fracture of the composite may be appreciable. Taking the work of fracture of the fibres as G_{fp} we have for the work of fracture of the composite $G_{1c} = V_fG_{fp} + V_mG_{mp}$.

5.3 ENERGY ABSORPTION BY CRACK-BRIDGING FIBRES

5.3.1 Fibre Pull-out

Fibre pull-out is associated mostly with brittle-matrix composites when work is done against frictional forces in extracting broken fibres from matrix crack faces. However, it can also occur in ductile matrix composites when fibre extraction is resisted by matrix shear.

When a load is applied to the end of a fibre, in order to extract it from a block of elastic material, the maximum interfacial shear stress occurs at the surface of the block. High interfacial shear stresses are similarly developed at the face of a matrix crack bridged by reinforcing fibres. If the matrix is ductile and the interface strongly bonded, shear displacements will occur in the matrix very near to the fibre surface so that a shear stress equal to the yield strength in shear of the matrix will be developed over the surface of an embedded fibre and resist its extraction from the matrix. If the matrix is brittle the high shear stresses developed at the matrix crack surface can cause failure of the interfacial bond and the debonded region can propagate down the fibre as an interfacial crack. Frictional contact may be preserved across the debonded interface and this in turn will be influenced by residual stresses produced by differential thermal contraction (both longitudinal and lateral) of fibre and matrix (Harris, 1978) and by Poisson's ratio effects (Takaku and Arridge, 1973; Kelly and Zweben, 1976). The latter authors point out that, although the Poisson's ratio effect will contribute to the debonding of the interface at the crack surface where the matrix is unloaded, this effect is also a function of the fibre volume fraction. The lateral expansion of the matrix in shedding load is restrained by the presence of adjacent fibres. Hence the fibres and matrix may still be held in contact. The fibre surface roughness is another significant factor controlling the residual frictional interaction between the debonded fibres and the matrix. Unless the residual interaction between fibres and matrix is effectively zero, the load required to extract a fibre will increase with increasing fibre length. At some particular embedded length, the stress developed in the protruding length of fibre will reach the fibre fracture stress. Fibres embedded by more than this length, therefore, will fracture instead of being extracted.

The shear strength of the frictional interface developed after debonding in conventional brittle-matrix composites (consisting, for example, of carbon, glass or boron fibres in a polymeric matrix) is dependent in some measure on thermal curing stresses and Poisson ratio effects. Because of the complexity of these interactions it is necessary to make some simplifying assumptions in developing an analysis of the physical processes occurring during the separation of the faces of a crack bridged by reinforcing fibres. As a first approximation, therefore, it is assumed that in brittle-matrix composites some specific value can be assigned to the fibre–matrix debonding energy and that following debonding any displacements occurring at the fibre–matrix interface take place at a constant frictional shear stress value. In the case of strongly bonded metal matrix systems we assume that debonding does not occur and the interfacial shear stress is taken as the yield strength in shear of the matrix.

Firstly we consider the behaviour of brittle fibres of uniform strengths and uniform lengths bridging a matrix crack. Work is done in developing strain energy in the fibre as it is loaded up prior to extraction. Also, as the fibre extends elastically prior to extraction, work is done against interfacial friction, or matrix shear, as a consequence of differential movement at the interface. These factors are small compared with the work done during pull-out and are not considered in the analysis given here. They are important in other circumstances and are included in the analysis given in Section 5.4 (ii).

When τ is constant, the stress carried by an embedded fibre diminishes linearly with increasing distance from the crack face. The stress σ required to be applied to the protruding fibre in order to extract it from the matrix will be given by $\sigma = 4\tau x/d$, where d is the fibre diameter and τ the shear strength of the interface. A critical value of $x = l_c/2$ will exist at which the ultimate tensile strength σ_{fu} of the fibre is just reached. Fibres that are embedded in the matrix by a shorter length than $l_c/2$ will be extracted, those whose embedded lengths are greater will fracture. We first consider that the length of fibre embedded in the matrix on one side of the crack face is x, where $x < l_c/2$. This embedded length of fibre will thus be extracted from the matrix as the crack faces separate. The force required to pull out the fibre will decrease progressively as the fibre is extracted. If the length of fibre embedded is x then the force required to extract it decreases linearly during extraction from $\pi d\tau x$ to zero. The displacement of the point of application of this force is x, so that the work done in extracting the fibre is $\pi d\tau x^2/2$. The fibres are all aligned with the direction of loading but are otherwise distributed randomly. Hence, for a matrix crack at any cross-section, the distribution of the embedded lengths will vary from zero to $l/2$, where l is the total fibre length. Thus the average work done in extracting one fibre is given by $\bar{w}_{fe}$ where

$$\bar{w}_{fe} = \frac{\pi d\tau}{2} \int_0^{l/2} \frac{x^2 \, dx}{(l/2)} = \frac{\pi d\tau l^2}{24} \tag{5.8}$$

Each fibre is extracted from one or other crack face so that only half of the fibres are extracted from each crack face. The effective surface energy γ_{fe} of one crack face due to fibre extraction from the matrix is therefore given by

$$\gamma_{fe} = \frac{V_f}{12d} \tau l^2 \tag{5.9}$$

If the embedded length is greater than $l_c/2$ the fibres will fracture instead of being extracted. Thus γ_{fe} (max) is given by

$$\gamma_{fe}(\max) = \frac{V_f d \sigma_{fu}^2}{48\tau} \tag{5.10}$$

We now consider the fibres all to have a length l where $l > l_c$. They have uniform strengths and, although parallel with each other, are axially distributed randomly within the composite. Fibres having one end within a distance $l_c/2$ of the matrix crack will be extracted and the others will fracture at the crack plane. Either end of a fibre may be extracted so that the fraction of fibres pulling out will be given by l_c/l. The distance x over which a fibre may be extracted varies between zero and $l_c/2$ and one-half of the fibres will be extracted from each crack face. The average work done $\bar{w}_{fe}$ in extracting a fibre is obtained by substituting l_c for l in equation (5.8) so that $\bar{w}_{fe}$ is now given by

$$\bar{w} = \left(\frac{l_c}{l}\right) \frac{\pi d \tau l_c^2}{24} \tag{5.11}$$

and the total work done per unit cross-sectional area of composite G_{fe} becomes

$$G_{fe} = \frac{V_f}{12} \left(\frac{l_c}{l}\right) \sigma_{fu} l_c \tag{5.12}$$

Substituting into equation (5.12) we have, for the effective surface energy of the composite,

$$\gamma_{fe} = \frac{V_f d \sigma_{fu}^2}{48\tau} \left(\frac{l_c}{l}\right) \tag{5.13}$$

which reduces to equation (5.10) when $l = l_c$. Thus, as has been pointed out by Kelly (1970), for the idealized conditions considered above, the effective surface energy of the crack face due to fibre pull-out increases in proportion to the square of the fibre length (equation (5.9)) up to a fibre critical length of l_c. As the fibre length increases above this value the effective surface energy of the crack faces, due to pull-out, diminishes with increasing fibre length and is inversely proportional to it (equation (5.13)). For the simple situation considered here, for which the interfacial shear stress is constant, only fibres embedded up to a distance $l_c/2$ are extracted. Also the force required to extract the fibres diminishes with increasing separation of the crack faces and becomes zero when the crack face separation becomes equal to $l_c/2$. Taking the crack tip to be approximately triangular in form (Figure 5.2), we see that the resistance of the fibres to the separation of the crack faces is developed over a limited region near the crack tip. The length of this zone will depend on l_c and the geometry of the crack tip.

If the interfacial shear strength between fibre and matrix is reduced considerably by an increase (over a critical value), in the tensile stress carried by the reinforcing fibre, the force required to extract a fibre can remain sensibly constant during fibre extraction and the pull-out length becomes very large. Bowling and Groves (1979) point out that this can occur during the extraction of ductile metal wires from an elastic matrix by the propagation of a yielded zone along the wire. Millman and Morley (1975) have described a similar process using high-strength tubular members. Interfacial decoupling can also be achieved by the elastic deformation of the reinforcing members. Morley and Millman (1974), Chappell et al. (1975) and Morley et al. (1976) have investigated a system in which the decoupling process is reversible. This mechanism allows the use of high-strength relatively brittle materials as reinforcing members and does not result in any significant reduction in the effective Young's modulus of the reinforcement. For these situations, the work done by fibre extraction is proportional to the area between the crack faces, so that the effective work of fracture continues to increase as the matrix crack increases in size.

5.3.2 Debonding and Elastic Stress Redistribution

In the simple analysis of the work done in extracting a conventional fibre from a matrix, described in §5.3.1, no account was taken of the need to overcome the adhesive bonding of the fibre–matrix interface, which is a necessary precursor to fibre extraction in the case of fibre-reinforced polymers. Interfacial debonding can occur if sufficient elastic energy is present in the system when the fibre is under load. Energy can be stored in the embedded length of fibre and also within a length of fibre protruding from the crack face. Energy is dissipated in debonding the fibre–matrix interface and by frictional losses occurring there. Chua and Piggott (1985) have developed a theoretical analysis of fibre pull-out taking these factors into consideration and Gray (1984) has given a review of various analytical treatments of fibre debonding and pull-out.

When the matrix is brittle, failure of the interfacial bond or cohesive failure of the matrix near the fibre–matrix interface will occur. This region of failure will initiate at the fracture face of the matrix where the interfacial shear stresses have a maximum value. The debonded region will then propagate along the fibre–matrix interface, providing sufficient work is done on the system to supply the surface energy of the newly created fracture faces. Generally, a residual frictional interaction will persist at the fibre–matrix interface after debonding. Its magnitude will depend on any residual differential stresses that may be present due to the thermal shrinkage of the matrix after curing and differential Poisson contraction effects. The surface roughness of the fibre may also induce mechanical keying. The work done per fibre in debonding the fibre–matrix interface over a distance x from the face of the matrix crack will be given simply by

$$2\pi d\gamma_{id}x = W_{id} \tag{5.14}$$

where γ_{id} is the surface energy of each of the two newly created fracture faces. The value of γ_{id} will depend on the degree of chemical bonding at the fibre–matrix interface. Since the fibres will generally be stronger than the matrix the upper limit of W_{id} will be set by the cohesive failure of the matrix, when γ_{id} will equal the surface energy of the matrix. In the case of a poorly bonded interface the value of W_{id} will approach zero. If it is assumed that after debonding the interface possesses a constant frictional shear strength τ, then the stress carried by the fibre at any distance x from the crack face will be given by $(\sigma - 4\tau x/d)$, where σ is the stress carried by the fibre at the crack face and d is the fibre diameter. The enhanced stress carried by a crack-bridging fibre produces an increase in fibre strain energy and the strain energy contained within a debonded length x of fibre is given by Kelly (1970) as

$$W_{fse} = \frac{\pi d^2}{8E_f} \int_0^x \left(\sigma - \frac{4\tau x}{d}\right)^2 dx = \frac{\pi d^2}{8E_f}\left(\sigma^2 x - \frac{4\sigma\tau x^2}{d} + \frac{16\tau^2 x^3}{3d^2}\right) \tag{5.15}$$

It is clear that W_{fse} has a maximum value when the fibre is on the point of fracturing. For this condition the maximum value of the debonded length is obtained when $x = d\sigma_{fu}/4\tau$ and for W_{fse} we have

$$W_{fse} = \frac{\pi d^2 \sigma_{fu}x}{24E_f} \tag{5.16}$$

This will occur if the fibre is discontinuous so that the stress carried by the end of the fibre is zero. In the case of continuous fibres a general stress σ_β must be carried by the fibres near the crack so that the maximum debonded length is now given by l_d where

$$d_d = d(\sigma_f - \sigma_\beta)/4\tau$$

Thus as the general stress carried by the composite outside the perturbed strain field near the crack approaches the failing stress of the fibres, the length of fibre which can debond and absorb additional strain energy prior to fibre fracture approaches zero.

After debonding, differential movement takes place between the fibre and the surrounding matrix and work will be done against residual frictional forces. Note that, if the matrix is ductile, work will be done in plastic deformation of the matrix in the vicinity of the fibre surface as the elastic strain distribution along the crack-bridging fibre under load is established. In the analysis that follows τ can be taken either as the shear strength of the debonded frictional interface of a brittle matrix composite or the yield strength in shear of a ductile matrix composite in which interfacial debonding does not occur.

We take the strain carried by the fibre where it emerges from the crack face to be ε_μ. Since τ is assumed constant the strain difference between fibre and matrix diminishes linearly with increasing distance from the crack face and becomes zero at the limit of the debonded length l_d (fibre stress transfer length). The strain difference between the fibre and the surrounding matrix at a distance x from the crack face is given by $\varepsilon_\mu(1 - x/l_d)$, so that the displacement of a point on the fibre distance x from the crack face with respect to the matrix is given by

$$dm_x = \int_x^{l_d} \varepsilon_\mu\left(1 - \frac{x}{l_d}\right)dx = \varepsilon_\mu\left(\frac{l_d}{2} - x + \frac{x^2}{2l_d}\right)$$

The work done by an element of fibre, length dx, against the interfacial shear force τ at this position will be given by $\pi d\tau\, dx\, dm_x$. The work done over the whole of the fibre matrix interface is therefore

$$W_{is} = \pi d\tau\varepsilon_\mu \int_0^{l_d} (l_d/2 - x + x^2/2l_d)\, dx$$

and

$$W_{is} = \frac{\pi d\tau\varepsilon_\mu}{6} l_d^2$$

or

$$W_{is} = \frac{\pi d\tau\sigma_{fu}}{6E_f} l_d^2 \tag{5.17}$$

If we assume $l_d = d\sigma_f/4\tau = l_c/2$, we have

$$W_{is} = \frac{\pi d^2\sigma_{fu}^2}{24E_f} l_d \tag{5.18}$$

The effective surface energy per unit area of crack face will be given by γ_{tot}, where

$$\gamma_{tot} = \frac{4V_f}{\pi d^2}(W_{id} + W_{fse} + W_{is})$$

and the work of fracture for $2\gamma_{tot}$. Experimentally fibre pull-out lengths and debonded lengths are found to be very variable. Estimated average values have been used in computing the work of fracture of a composite (Kirk *et al.*, 1978).

The maximum stress carried by a debonded fibre in residual frictional contact with the matrix is developed at the crack faces, so that a fibre having a uniform strength will fracture at that position. The fractured fibre ends will then contract back into the matrix and the strain energy contained in the fibres when fracture occurs will be dissipated in frictional losses and by the propagation of elastic waves. If the fibres have a distribution of flaws along their lengths, fibre fracture may occur at a point some distance from the crack face. The strain energy contained in the fibre at fracture will be dissipated as before, but further work will be done in extracting the broken end of the fibre against residual interfacial frictional forces. The amount of work done during pull-out under these conditions depends on the position at which the fibre fractures. The load supported by the crack bridging fibre will also depend on the position and severity of the flaw. In many applications fibres are incorporated into a matrix in the form of bundles. These act as a single reinforcing member whose effective size depends on the number of fibres present but which is typically larger by one or two orders of magnitude than the diameters of the individual fibres. The tensile strength of a fibre bundle is more consistent than that of the individual fibres within it and failure of fibre bundles takes place quite near to the crack face.

The relative importance of fibre pull-out as an energy-absorbing process, compared with the work done during the debonding, elastic extension and subsequent fracture of a continuous crack-bridging fibre, depends strongly on the ratio between the length of fibre debonded and the length extracted. If these are comparable, the work done in extracting a fibre will be an order of magnitude greater than the work expended in the other processes. (For a discussion of this see Kelly, 1970.)

In many practical systems the lengths of fibres extracted from the matrix are much less than the distances over which the fibres are observed to become debonded from the matrix. The length of the debonded zone is easily measured in the case of glass-fibre-reinforced materials, but is not easily observed in the case of carbon-fibre composites. Measurements can be readily made of the lengths of fibres extracted. The relative importance of the various energy absorbing mechanisms can be calculated for any particular system from measurements of the work of fracture, fibre pull-out lengths and other material properties. Wells and Beaumont (1982) have used this type of analysis to predict the effect of changes in material parameters on the work of fracture of a composite system.

In appropriate circumstances, the work done in extracting a fibre can be much greater than the work expended in debonding and stressing to failure crack-bridging fibres that do not pull out of the matrix. However, fibre pull-out necessitates that the crack faces separate by an appreciable amount. This may be acceptable in some applications but may cause undesirable effects in others.

Double-cantilever beam test specimens provide a convenient means of investigating the influence of crack-bridging fibres on the propagation of a crack in a polymeric matrix (Figure 5.1). The changes in the strain distribution following fibre debonding are readily observable using polarized light, thus enabling the debonded length of fibre to be measured.

Investigations have been made of the behaviour of single crack-bridging glass fibres (Harris *et al.*, 1975; Harris and Ankara, 1978) and of glass- and carbon-fibre bundles (yarns) of varying degrees of twist (McGarry and Mandell, 1972). The debonding, elastic extension and eventual failure of the crack-bridging reinforcing members was observed directly. The debonded length is generally much greater than the length of fibre eventually extracted from the matrix. It is observed that, in the case of yarns, the debonding process depends on the degree of twist imparted to the fibre bundle. This influences the extent to which the yarn debonds as a unit instead of behaving as a bundle of individual fibres. Also, when a number of yarns are twisted together, the debonding process is observed to depend upon the geometrical arrangement of the bundle. In cases where the individual twisted yarns fit closely together, presenting a smooth exterior, debonding is facilitated. When the surface of the assembly is irregular debonding is inhibited, as would be expected from mechanical keying effects with the polymeric matrix.

5.3.3   Oblique Fibres and More Complex Fibre Assemblies

The effect of the orientation of the fibres to the direction of crack growth is of particular significance in the behaviour of fibrous laminates. Studies have been made, using a double-cantilever beam specimen, of the behaviour of fibre yarns oriented at $\pm \theta °$ to the normal crack direction (McGarry and Mandell, 1972). The fibre yarns crossed at the position of the propagating crack so as to give a balanced system. No significant effects were apparent from orientation changes over the range 0° to 60° and the debonded length remained approximately constant for all orientations.

Double-beam cantilever specimens have also been used by McGarry and Mandell (1972) to investigate the mechansim of crack growth in cross-ply laminates. In these specimens the fibres in the longitudinal plies were arranged to be perpendicular to the primary crack with those in the transverse plies parallel with the primary crack. As a load is applied to the ends of the cantilever beams, splits are formed at the tip of the primary crack within the longitudinal plies and parallel to the longitudinal fibres. These splits are thus perpendicular to the primary crack which, as a consequence, becomes stabilized. The propagation of the splits away from the primary crack, by a process of forward shear, is resisted by the fibres of the transverse plies, which bridge the splits and transfer shear loads across them. In this way tensile loads are applied to the longitudinal plies beyond the first split. This load transfer process eventually causes failure of the fibres in the longitudinal plies beyond the position of the first splits and the primary crack propagates in a stepwise fashion becoming stabilized by further splits. The process repeats itself as the crack grows across the laminate. The behaviour of a laminate is thus broadly similar to that of a model system containing individual crack-bridging elements but is found to be influenced by the precise nature of the ply stacking sequence. Similar splits occur in woven-fabric-reinforced systems and the crack growth mechanism is similar to that of a laminate construction. Composite systems reinforced by a random arrangement of individual fibre bundles behave in a similar manner, in that debonded bundles bridge and restrain the crack tip. This leads to the development of a general zone of damage due to multiple cracking around the nominal position of the crack tip.

The strength of obliquely bridging ductile fibres is maintained because of plastic flexure of the fibre at the crack face. Frictional effects are enhanced owing to the high local pressures exerted at the interface near the plastic hinge and the fibre will be with drawn from the matrix if its embedded length is sufficiently short. During this process successive portions of a ductile fibre are deformed with the absorption of energy (Morton and Groves, 1974).

5.4   THE MECHANISM OF TENSILE FRACTURE IN UNIDIRECTIONALLY REINFORCED COMPOSITES

Unidirectional composites consist of a bundle of brittle fibres aligned parallel with each other and encapsulated in a matrix that has a much lower elastic modulus and strength than the fibres. In practice the fibres are not uniformly dispersed or perfectly parallel with each other. The fibres themselves exhibit considerable variability in strength owing to flaws that vary in severity and these are distributed randomly along the length of individual fibres.

Because of the wide variablity in fibre strengths, some of the fibres fracture before the composite reaches its failing strain. The analysis of tensile fracture therefore requires an understanding of the processes by which the failure of individual fibres or groups of fibres leads to the failure

of the composite. A comprehensive analysis must also accommodate the
known insensitivity of unidirectional composites to surface damage, notches
and holes.

5.4.1 Fracture as a Consequence of Cumulative Fibre Failure

The cumulative failure analysis rests on the argument that fibre failures will
be distributed throughout the composite and that their numbers will
increase as the load applied to the composite increases. The load-bearing
ability of a particular fibre is lost at the point of fracture but this loss of
strength is localized because stress is transferred from the matrix to the fibre
over the fibre–matrix interface near the point of fracture. Thus at some
distance from the point of fracture (the stress transfer distance) the
load-bearing ability of the fibre is unimpaired. This stress transfer process
increases the load carried by fibres adjacent to the broken fibre and thus
increases the probability of sequential fibre failure. These effects depend on
the degree of coupling between the fibres and the matrix. At one extreme
we can assume zero coupling, so that the assembly behaves as a bundle of
isolated fibres. Under these conditions the load carried by a fibre before
fracture is distributed uniformly amongst the still-intact fibres. The failing
strength of the fibre bundle can be calculated from the known distribution
of fibre strengths (Coleman, 1958). The load-bearing efficiency of the
bundle, measured as the ratio between the strength of the bundle and the
average strength of the fibres contained within it, decreases as the variability
in the fibre strength increases. This is illustrated in Figure 5.3, where the

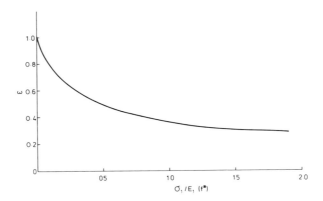

Figure 5.3. The strength efficiency ε for a bundle of fibres versus the coefficient
of variation $\sigma/E(f^*)$ in the strength of the component filaments. (After Coleman,
1958.)

bundle strength efficiency is plotted against the coefficient of variation in fibre strengths (standard deviation divided by the mean). When the fibres and matrix are strongly bonded together the composite behaves as a homogeneous brittle solid and its strength is governed essentially by the size of the flaws that may be present and the surface energy of the material (see Chapter 1, §1.1). For intermediate degrees of coupling the analysis becomes difficult and various assumptions have to be made in order to deal with the problem (see Harlow and Phoenix, 1978 for a review).

5.4.2   Chain of Fibre Bundles Model

The composite is regarded as being cut into slices to form a chain of bundles. Failure occurs when one bundle reaches its failing strain. The length of each bundle is defined by the length of fibre over which the local strain field is perturbed by the fracture of a single fibre. The fibres which are the immediate lateral neighbours to the broken fibre carry most of the load previously carried by the broken fibre and it is assumed that fibres at increasing distances from the site of initial fibre fracture bear a decreasing proportion of the extra load. It is difficult to calculate stress enhancement factors for all configurations of fractured and intact fibres and various approximate analytical models have been proposed by use of which it is possible to calculate the progressive failure of fibres from assumptions of their initial strength distributions. Attempts have also been made to compute the effects of various clusters of broken fibres (Smith *et al.*, 1983). In this way the predicted presence of various numbers of adjacent fractured fibres can be correlated with the experimentally observed strength of the relevant composites. Zweben and Rosen (1970) proposed that the fracture nuclei contain between two and four broken fibres. Barry (1978) calculated that failure groups containing five or six broken fibres could occur before failure in a particular carbon-fibre–epoxy resin system, and a similar criterion has been proposed for composite failure by Harlow and Phoenix (1981).

5.4.3   Crack Growth Models

It has been suggested by Batdorf (1982) that a group of adjacent fractured fibres can be regarded as a crack that can propagate at some critical stress level by the sequential failure of surrounding fibres. The Batdorf model proposes that the stress transfer length associated with a group of broken fibres should be a function of the number of fractured fibres in the group. Account is also taken of the non-uniform spatial distribution of fibres in real composites, which leads to a non-uniform redistribution of load from the fractured fibres to the surrounding intact fibres (Batdorf and Ghaffarian, 1982). The probability of a group of i broken fibres being present increases as the load carried by the composite increases. Also the load required to produce i adjacent broken fibres increases as i increases. Composite failure is assumed to occur when the stress applied to the composite is sufficient to ensure that at least one unstable group of fractured fibres is present in the composite structure.

A mechanism for the growth of a matrix crack, leading eventually to the failure of the composite, has been put forward by Morley (1983). Matrix crack growth is assumed to occur from a flaw of arbitrary size within which the fibres have fractured. As the matrix crack propagates it encounters and by-passes intact fibres which then bridge the crack. Calculations made on the basis of the model show that the crack will eventually become stable as a consequence of the effects of crack-bridging fibres even though its length has increased. Stable cracks of increasing length are formed as the load applied to the composite is increased. Eventually the additional stress carried by the crack-bridging fibres becomes great enough to cause the weaker ones to fail. At some critical stress level, which will depend upon the strength distribution of the fibres and the other mechanical parameters of the composite, the crack can become unstable and propagate by reason of the sequential failure of the still-intact crack-bridging fibres. This analysis can be applied to an idealized 2-dimensional layer of fibres and also to penny-shaped cracks.

An alternative failure process based on this model envisages the growth of large numbers of stable cracks within the composite structure. At some particular stress level these become linked by shear failure in the matrix, thus causing catastrophic failure of the composite. Fractography studies carried out by Purslow (1981) show that carbon-fibre-reinforced plastics exhibit relatively planar failure when the fibres are strongly bonded to the matrix and that sequential fibre fracture can be inferred. Also small groups of fibres appear to fracture independently of each other on slightly different planes prior to ultimate failure of the composite. Further information on progressive fibre failure during loading in carbon-fibre–polymeric-matrix composites can be obtained from carbon-fibre–epoxy composites by removing the matrix from around the surface fibre layers with sulphuric acid (Fuwa et al., 1975). Various types of fibre failure can be observed by the use of this technique. The failure of small groups of fibres in an associated manner which become linked in a more or less planar fracture surface by shear failures in the matrix, as previously noted, are observed. The existence of associated fibre fracture on planes not normal to the fibre alignment has also been observed. A general distribution of individual fibre failures is also present.

The failure process of unidirectional fibrous composites containing fibres of non-uniform strength is complex and not fully understood, and the experimental variability in tensile strengths presents a major problem with regard to engineering design. In practical applications the certainty of some degree of surface damage with the localized failure of fibres extending to some distance below the surface of a component has to be taken into account in evaluating the damage tolerance of fibre composite systems. Techniques for the analysis of the effects of relatively small amounts of surface damage in carbon-fibre–epoxy-resin composites have been proposed by Morley (1985a).

5.5   FRACTURE OF HYBRID FIBRE COMPOSITES

By using fibres of more than one type to reinforce a matrix a variety of composite systems can be produced. When the two fibres used differ markedly in cost this approach is particularly useful in producing materials that have both a range of mechanical properties and a range of material cost values. In other situations the inclusion of fibres having a high extension at failure has been observed to enhance the contribution of low-extension fibres to the strength and stiffness of a composite.

We have previously noted that the dry bundle strength of flawed fibres is modified when the fibres are bonded together and is controlled by the properties of the fibre matrix interface and by the fibre and matrix elastic properties. The situation is further complicated when fibres of more than one type are present since not only do the fibres have different flaw distributions, mechanical properties and perhaps also different interfacial characteristics, but they will generally also be mixed together in a random manner. Thus a fibre of one type may be surrounded by nearest neighbours of the same type or by various combinations of the two types of fibres present in the composite system (Parratt and Potter, 1980).

The lower bound of composite failure can be assumed to correspond to the failure of two adjacent fibres at the same position. Zweben (1977) has examined this condition for an array of alternatively positioned low-failure-extension and high-failure-extension fibres. Failure at a low-extension fibre increases the local stress carried by the adjacent high-extension fibre as calculated from shear lag analysis. The stress distribution so calculated is replaced by an equivalent step-function giving the same average enhanced fibre strain, which is in turn used to calculate the increased probability of localized fibre failure. The analysis predicts that a hybrid composite containing high-extension fibres will fail at a higher strain than a composite containing only low-extension fibres. The theory is qualitatively correct since it is observed that hybrids do possess higher failing strains than composites constructed only from low-failing strain fibres (Bunsell and Harris, 1974). However, the numerical predictions are only in rough agreement with the experimental data.

Aveston and Sillwood (1976) have used a modified form of the energy-balance analysis given in §4.3 to deal with the behaviour of a glass-plus carbon-fibre-reinforced polymeric matrix. Both fully bonded (shear lag) and debonded (constant frictional shear strength interface) conditions were considered in the analysis. In the case of these studies the carbon fibres formed only about 10% of the total reinforcement but their presence was obvious from the enhanced elastic modulus of the composite compared with a wholly glass-fibre-reinforced system, which had a much greater failing strain. An inflexion is observed in the stress–strain curve. The failing strain of the carbon fibres, deduced from the position of the inflexion in the load–extension curve, is much greater than the inherent failing strain of the carbon fibres themselves. These effects can be accounted for by a mechanism that inhibits the growth of cracks originated from fractured carbon fibres so that the low-failing-strain fibres continue to contribute to the elastic modulus of the composite as they become fractured into progressively shorter lengths (Morley, 1985b). Experimental support for this view comes from studies carried out by Wadsworth and Spilling (1968), who observed that carbon fibres bonded to a massive aluminium plate by a thin epoxy resin layer were broken into many fragments at their normal failing strains as the aluminium plate was extended. The stress transfer from the broken set of fibres to the polymeric matrix seems to occur by frictional interfacial slipping.

When hybrid composite tapes consisting of various proportions of glass fibre and carbon fibre strands are deflected in bending, so as to give a controlled fracture, the debonded length of the glass fibres and the lengths of fibres of both types extracted from the fracture faces can be observed, Kirk et al., (1978). The debonded lengths of the glass fibres are found to be typically about twenty times greater than their pull-out lengths. For the system examined, the carbon fibres contribute to the work of fracture of the hybrid primarily by the pull-out mechanism whereas energy absorption by frictional losses prior to pull-out was more important in the case of the glass fibres. The mechanisms of cracking and failure in hybrid composites and laminates have been reviewed by Aveston and Kelly, (1980).

5.6 CONCLUSIONS

The relevance of the various energy-absorbing processes to the fracture toughness of a fibre composite depends on the engineering conditions pertaining to the fracture process. Large amounts of energy can be absorbed by fibre pull-out but this proceeds at a diminishing stress level. By contrast, debonding, stress redistribution and post-debond frictional losses can proceed at an increase of applied load and because of this may be of greater importance in some engineering circumstances. The issue is complicated by the inherently non-homogeneous nature of the microstructure of fibrous composites and the variety of failure processes that are possible.

6
Failure under Off-axis and Compressive Loading

As discussed in Chapter 3, the elastic properties of fibre composites and laminates can be calculated with reasonable precision from a knowledge of the properties of the fibres and the matrix using idealized models of the composite structure. This approach is feasible because the elastic properties manifest themselves as values which are averaged over the total volume of the composite, so that localized inhomogeneities are not of major significance. In contrast, local inhomogeneities or flaws control the strength of fibrous composites, since they initiate failure. Flaws may be points of weakness in the fibres themselves or may take the form of unavoidable localized variations in the composite structure such as matrix voids or points of fibre-to-fibre contact. The strength of the composite is controlled by the stress levels at which these flaws propagate. Composite failure may occur as a consequence of fibre failure, matrix failure or failure of the fibre–matrix interface. The form of failure depends on the relationship between the fibre orientation and the direction of the applied load. Shear forces applied parallel to the direction of fibre alignment will tend to produce matrix failure or failure of the interface in longitudinal shear. Transverse tensile loads cause failure of the matrix or interface. On the other hand, tensile loads applied parallel to the fibre direction produce failure by fibre fracture. Compressive loads, applied in the direction of fibre alignment, can cause failure through localized buckling of the fibres, usually due to inadequate local support by the matrix. Compressive loads applied in other directions can cause shear failure of the matrix. Under complex loading conditions one or more failure processes may be occurring simultaneously and these may interact with each other and so modify the stress level at which failure occurs.

Because of the highly anisotropic nature of fibrous composites the stress levels at which the pre-existing flaws propagate within a lamina are very dependent on the relative orientation of the fibres to the direction of loading. In the case of multi-ply laminates, crack extension may take place in plies orientated in particular directions and not in others, so that the laminate may retain a load-bearing capability but have a reduced elastic modulus.

There are very few practical applications in which unidirectionally reinforced composites are subjected only to tensile loads. Unidirectional systems may be loaded in flexure, and then tensile, compressive and shear strengths become important. Laminates are usually subjected to combinations of in-plane compressive, tensile and shear stresses of different magnitudes and may, in addition, be subjected to flexural stresses. The tensile strength of a lamina measured in the direction of the fibre alignment is much higher than its transverse tensile strength or its in-plane shear strength. Uniaxial compressive strengths may be higher or lower than the corresponding tensile strengths. When a combination of loads is applied, both the failure stress and the actual mechanism of failure depend on the loading conditions. Furthermore, failure processes occurring in individual laminae can be inhibited by adjacent laminae when laminae are bonded together to form laminates. As discussed in §4.6, this effect depends on the thicknesses of the laminae. It also depends on the orientation of the fibres in adjacent laminae.

In this chapter the transverse tensile, compressive and interlaminar shear strengths of unidirectional composites are first discussed. The effect of off-axis tensile loading on the strength of a unidirectional lamina is then examined and this leads to an account of the failure processes taking place in laminates. We then deal with the effect of combined stresses on failure processes.

6.1 TRANSVERSE TENSILE FAILURE OF UNIDIRECTIONAL COMPOSITES

The transverse tensile strength and failing strain of a unidirectional fibre-reinforced polymer is very low. Typical values for carbon fibres and glass fibre reinforced polymers containing respectively 62% and 55% by volume of fibres are shown in Table 6.1. In fibre-reinforced metals, transverse strengths and elastic modulus values are an order of magnitude

TABLE 6.1
Mechanical properties of unidirectional laminates
(Bailey *et al.*, 1979)

	0° CFRP*	0° GFRP†	90° CFRP	90° GFRP
Young's modulus (GN m^{-2})	127	42	8.3	14
Fracture stress (GN m^{-2})	1.7	0.92	0.039	0.056
Fracture strain (%)	1.16	2.2	0.48	0.5

* Carbon-fibre-reinforced polymer.
† Glass-fibre-reinforced polymer.

greater than those of polymeric matrix composites. The low transverse strengths and failing strains of polymeric matrix composites presents a major problem in the design of laminates, since there are few engineering applications where transverse stresses are not encountered. Initial failure usually occurs by the propagation of cracks parallel with the fibres in laminae within which appreciable transverse strains are developed. Generally, the transverse strength of a polymeric matrix composite is much less than that of the unreinforced matrix and transverse tensile failure is observed to be initiated in regions of high local fibre concentration. Transverse cracks propagate preferentially in these regions. The transverse tensile strength is governed by a number of factors such as the degree of adhesion between the fibres and the matrix, the presence of voids, and the elastic and plastic characteristics of the fibres and matrix.

A simple analysis, first developed by Kies (1962) illustrates the reason for the low transverse strengths of polymeric matrix fibre composites. An idealized cross-section of a composite with the fibres in a square array is illustrated in Figure 6.1. The composite is assumed to be carrying an average transverse tensile strain of $\bar{\varepsilon}_x$. Because the fibres have a very much higher elastic modulus than the polymeric matrix, their transverse extension under the applied transverse stress is much less than that of the matrix. Hence, the portions of matrix within the zone shown shaded in Figure 6.1 experience an enhanced strain ε_m compared with the matrix in a similar zone that passes between the fibres. According to Kies the strain magnification within the resin along a line joining the fibre centres is given by

$$\left[\frac{\varepsilon_m}{\bar{\varepsilon}_x}\right] = \left[2 + \frac{s}{r}\right] \bigg/ \left[\frac{s}{r} + 2\left(\frac{E_m}{E_f}\right)\right] \tag{6.1}$$

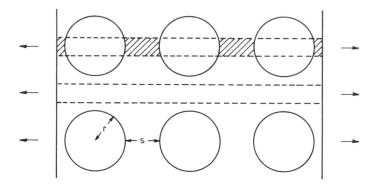

Figure 6.1. Schematic representation of matrix strain enhancement in a unidirectional lamina subjected to a transverse tensile load.

where s is the distance between the fibres and $2r$ the fibre diameter (see Figure 6.1). This equation illustrates the effect on the strain enhancement of the relative elastic modulus of matrix and fibres (E_m and E_f) and the distance between the fibres.

More accurate stress analyses have been carried out (see e.g. Adams and Doner, 1967b; Tirosh *et al.*, 1979). The use of photoelastic techniques in the analysis of stress distributions has been described by Daniel (1974). Composite transverse tensile strengths are difficult to predict from analytical studies because of the possibility of plastic deformation in the matrix, together with the effects of residual thermal stresses, and also because of the non-uniformity of fibre distribution and the presence of weak interfaces and matrix voids and flaws. Generally, the transverse strength and failing strain of a composite are much lower than those of the matrix itself. However, the composite transverse failing strain values are increased if the matrix has a high failing strain (Legg and Hull, 1982).

Some degree of fibre misalignment occurs in practical fibre composites, so that a crack propagating in the nominal direction of fibre alignment eventually becomes bridged obliquely by misaligned fibres. As the crack faces separate, energy is absorbed as the fibres are torn out of the matrix. When the crack face separation becomes large enough, the oblique crack-bridging fibres fracture and make no further contribution to energy absorption. Thus when the crack is short, the work of fracture corresponds to that of the matrix or fibre–matrix interface. As the crack extends, the work of fracture increases owing to the effect of the oblique crack-bridging fibres and reaches a maximum value when the fibres behind the advancing crack tip start to fracture sequentially as the crack advances. The effective work of fracture depends on the crack tip geometry since the separation of the crack faces controls the size of the zone behind the crack that is bridged obliquely by fibres. This effect has been confirmed experimentally by Phillips and Wells (1982) using a double-cantilever beam technique (Figure 5.1(*b*)). The apparent work of fracture during crack growth increases as the beam compliance, and hence crack face separation, is increased.

6.2 FAILURE IN LONGITUDINAL COMPRESSION

The strengths and elastic moduli of unidirectionally reinforced polymeric matrix composites loaded in compression in the direction of fibre alignment are found to depend upon a large number of factors. The measured strengths of composites are very dependent on the uniformity of fibre distribution, on the presence of matrix voids and on the details of the experimental techniques used. Other factors of importance include the matrix yield strength and the effectiveness of the bonding between the fibres and the matrix.

The strengths of a variety of experimental samples, constructed under laboratory conditions and tested in a simple manner, have been investigated by Piggott and co-workers (Piggott and Harris, 1980, 1981; Piggott and Wilde, 1980; Martinez *et al.*, 1981). Although the experimental arrangements produced relatively low compressive strength values, the study illustrates many of the important factors governing the characteristics of fibre composites loaded in uniaxial compression. The elastic modulus follows a rule of mixtures relationship, particularly at low stress levels. The strengths are also generally proportional to the fibre content up to particular fibre volume fractions and stress levels, but this linear relationship may not be preserved at high fibre contents and high stress levels. Typical examples of the experimental relationship between compressive strength and fibre volume fraction are shown in Figure 6.2. Composites utilizing fibres that have low intrinsic compressive strengths, such as Kevlar aromatic polyamide fibres, are also weak in compression. At the relatively low fibre volume fraction of 0.3 the fibre distribution is not particularly uniform and the composite compressive strengths are relatively low and are not sensitive to fibre straightness and fibre misalignment. Indeed, there is some evidence that for composites of this type the compressive strengths are improved as the fibre misalignment is increased up to about 10° (Martinez *et al.*, 1981).

Composite longitudinal compressive strengths σ_{lc} are known to fall with decreasing matrix yield strength values and can be represented by a power-law relationship of the form

$$\sigma_{lc} = A(\sigma_{my})^n \tag{6.2}$$

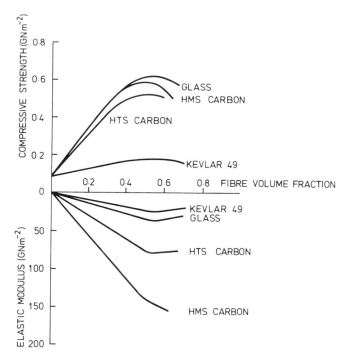

Figure 6.2. Uniaxial compressive strengths and elastic moduli for a polyester resin reinforced with various fibres. (Redrawn from Martinez *et al.*, 1981.)

for a wide range of matrix yield strength values σ_{my} (Piggott and Harris, 1980). The strengths of polymeric matrix composites fall with increasing temperature; this can also be ascribed to the reduction in the matrix yield strength that occurs under these conditions. Experimental studies of hybrid systems, in which the fibres are distributed fairly irregularly, show the compressive strengths and elastic moduli to correspond approximately to a rule of mixtures relationship. There is some experimental evidence that strengths can be reduced below the rule of mixtures value by cohibitive effects. Even greater reductions below the rule of mixtures values are observed for compressive elastic modulus values. Piggott (1981) has suggested that composite failure in compression can be brought about by a number of mechanisms, so that the compressive strength will be dictated by whichever process operates at the lowest stress level. The following factors are likely to be significant: fibre strength and initial curvature, matrix characteristics, and fibre–matrix adhesion.

Much higher compressive strengths than those illustrated in Figure 6.2

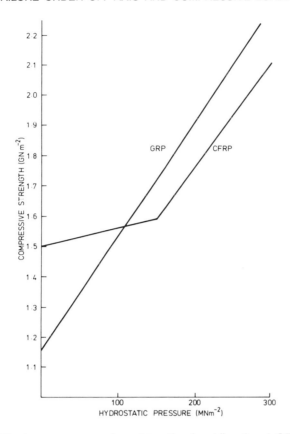

Figure 6.3. Maximum compressive strength of unidirectional CFRP and GRP specimens tested under a superposed hydrostatic pressure. It is postulated that at hydrostatic pressures $<150\,\mathrm{MN\,m^{-2}}$ CFRP specimens fail by the buckling of fibre bundles. At higher pressures the composite strength is controlled by matrix yielding. (Redrawn from Wronski and Parry, 1982.)

are observed with carefully manufactured specimens of polymer matrix composites containing a high volume fraction of glass or carbon fibres. Wronski and Parry (1982) and Parry and Wronski (1982) observed that the strength is increased further by the addition of a superposed hydrostatic pressure; this effect is illustrated for glass fibre and carbon fibre–epoxy resin composites in Figure 6.3. Since a superposed hydrostatic pressure affects different failure mechanisms in different ways, the technique provides important information on the mechanics of compressive failure. The experimental samples utilized by Wronski and Parry were produced from pultruded rod containing 60% by volume of carbon or glass fibres in

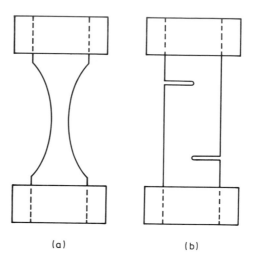

(a)   (b)

Figure 6.4. Illustrating test specimens used by Parry and Wronski. (Redrawn from Parry and Wronski, 1982.)

an epoxy resin matrix. Some samples were machined to form a reduced section (Figure 6.4(a)); others contained two saw cuts (Figure 6.4(b)) and provided data on shear strengths under a superposed hydrostatic pressure.

The characteristic failure mode of unidirectional fibre-reinforced specimens is by a "kinking collapse" or compressive creasing. It is illustrated in Figure 6.5 and has been observed in various experimental situations (Parry and Wronski, 1982; Wronski and Parry, 1982; Argon, 1972; Chaplin, 1977; Weaver and Williams, 1975). The compressive crease is initiated by localized buckling of the fibres; this then extends across the specimen as a band of fractured fibres at an angle of about 30° to the cross-section of the composite. Compressive failure is catastrophic but by encapsulation of the specimen in a transparent resin the load can be removed and the specimen examined at various stages during the failure process. By this means, fibre fracture ahead of the kink band can be observed, the fibres fracturing in a plane perpendicular to the axis of loading (Parry & Wronski, 1982). Wronski and Parry (1982) observed that for carbon-fibre-reinforced polymer specimens the widths of the kink bands were not affected by the magnitude of the superposed hydrostatic pressure and were fairly consistently within the range of 250 μm to 500 μm. At a superposed hydrostatic pressure of less than $150\,\mathrm{GN\,m^{-2}}$, longitudinal splitting occurs from the reduced cross-section of carbon-fibre-reinforced polymer (Figure 6.4(a)) prior to failure by kink band propagation. The compressive strengths of the

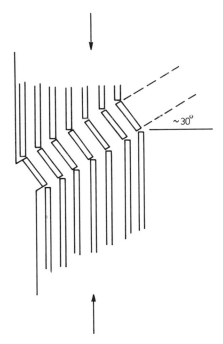

Figure 6.5. Illustrating failure in longitudinal compression by the propagation of a kink band.

specimens increase rapidly with increasing hydrostatic pressure above $150 \, GN \, m^{-2}$ (Figure 6.3). This contrasts with glass-fibre reinforced polymer specimens, in which a uniform increase in the compressive strength occurs with increasing hydrostatic pressure. Wronski and Parry have suggested that this increase in strength is due to the increase in the yield strength of the polymer matrix, which shows a similar dependence on hydrostatic pressure. The same argument can be applied to the behaviour of carbon-fibre-reinforced polymer specimens at high superposed hydrostatic pressure values.

In-plane shear strengths, measured by use of specimens of the type shown in Figure 6.4(b), are also observed to increase with increasing values of superposed hydrostatic pressure for both carbon-fibre and glass-fibre-reinforced composites (Figure 6.6). Strengths are calculated on the basis of the nominal minimum area supporting in-plane shear. In the case of carbon-fibre-reinforced composites, shear failure occurs only for superposed hydrostatic pressures less than $150 \, MN \, m^{-2}$. At higher superposed pressures, failure occurs by the propagation of a kink band from the tip of

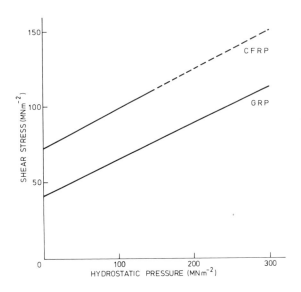

Figure 6.6. Illustrating the effect of superposed hydrostatic pressure on the apparent interlaminar shear strength of CFRP and GRP specimens. Kink band failure is illustrated by dashed line. (Redrawn from Wronski and Parry, 1982.)

the machined notches. This phenomenon is not observed for glass-reinforced polymer specimens; for this material failure is by mid-plane shear at all superposed hydrostatic pressures investigated.

A theoretical analysis of the compressive failure of a unidirectional composite was first proposed by Rosen (1965). This analysis was based on the assumption of the cooperative buckling of the fibres throughout the composite. On this basis, the compressive strength σl_c is given by

$$\sigma_{lc} = G_m/(1 - V_f) \tag{6.3}$$

where G_m is the shear modulus of the matrix. For realistic fibre volume fractions this analysis predicts compressive strengths of several giganewtons per square metre for a polymeric matrix composite and much higher values for metal matrix systems. Experimentally, equation (6.3) is found to overestimate the compressive strengths of fibre-reinforced plastics and metals and it does not predict the observed relationship between compressive strength and fibre volume fraction. It assumes uniform buckling of the fibres throughout the structure and can be regarded as an upper theoretical bound that is unattainable in practice owing to localized failure from flaws in the composite structure.

It is argued by Wronski and Parry (1982) that the experimental observations of the effect of superposed hydrostatic pressure on the compressive strength of carbon and glass-fibre-reinforced composites preclude any failure process involving shear failure of the fibres. If this were so the slope of the plot of compressive strength hydrostatic pressure would be unity instead of the observed values which are in excess of 3. (Figure 6.3). Furthermore, the inclination of the fracture surfaces would be expected to be 45° instead of the observed values of about 30°. Theoretical analyses of compressure failure due to local fibre buckling have been put forward by Piggott (1981) and by Wronski and Parry (1982).

6.3 TRANSVERSE COMPRESSIVE STRENGTH

The transverse compressive strength of a fibrous composite is of engineering importance in the design of wedge joints, which provide a means of transferring loads from a metal component to a composite material. An important example is the use of dovetail joints as a means of attaching fibre-reinforced composite compressor blades to a hub assembly. In such applications the load applied to the compressor blade is primarily tensile due to high centrifugal forces, but compressive loads are developed within the wedge joint. It is therefore of interest to examine the strengths and modes of failure of unidirectional fibre-reinforced composites subjected to transverse compressive loading.

In a homogeneous material, shear failure due to a compressive load would be expected to occur on the plane of maximum shear, i.e. at an angle of 45° to the loading direction. In a unidirectional fibre-reinforced system, the matrix may fail on planes parallel to the fibres and this does not require that the fibres should fail. Failure on planes containing the fibres necessitates fibre failure. When a compressive load is applied only in the 2-direction in carbon-fibre-reinforced epoxy resin composites (Figure 6.7) failure takes place on planes parallel to the fibres and the composite strengths are insensitive to the fibre volume fraction (Figure 6.8). The strengths measured under these conditions are higher than the in-plane shear strengths (see §6.5) because of the presence of a component of compressive stress acting perpendicularly to the plane of shear failure.

When a constraint is applied in the 3-direction (Figure 6.7) the strengths are very much increased (Figure 6.8) and failure is observed to occur on planes parallel to the direction of constraint, so that fibre fracture then occurs. For these loading conditions the strengths are dependent on the fibre volume fraction, which is consistent with the assumption that the fibre shear strength is much greater than that of the polymeric matrix. The

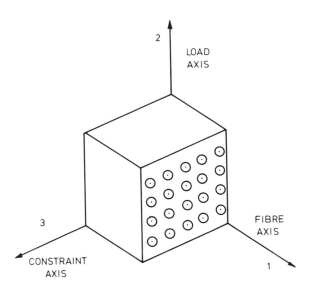

Figure 6.7. Illustrating the relationship between the fibre orientation and the directions of the compressive load and constraint applied to the specimen. (After Collings, 1974; based on British Crown copyright/RAE original figures.)

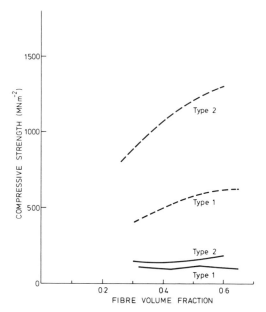

Figure 6.8. Transverse compressive strengths of unidirectional Type I and Type 2 carbon-fibre–epoxy-resin composites. Effect on transverse constraint shown as (– – – –). (After Collings, 1974.)

higher strengths observed for composites constructed from Type 2 carbon fibres implies that these fibres have higher shear strengths than Type 1 fibres.

6.4 FAILURE IN FLEXURE

Many types of composite structures are subjected to flexural deformations and thus have to withstand a combination of tensile, compressive and shear stresses. It is convenient to examine the characteristics of a composite material under these conditions by subjecting a short unidirectionally reinforced beam to 3-point bending (Figure 6.9). Using elementary beam theory and neglecting any effects due to concentrated loads, the maximum tensile and compressive stresses σ are given by

$$\sigma = 3PL/2wt^2 \qquad (6.4)$$

where P is the applied load, L is the span, w is the width and t is the depth of the beam. The maximum shear stress occurs at the neutral axis and is given by

$$\tau = 3P/4wt \qquad (6.5)$$

Examination of the fracture surfaces of beams loaded in flexure shows compressive and tensile failure zones meeting at approximately the neutral axis. This is illustrated in Figure 6.10(a). The compressive zones have a stepped surface and higher magnification shows that the fracture face of the individual fibres have two zones apparently caused by tensile and compressive failure in buckling (Figure 6.10(b)).

The 3-point bending test has often been used to evaluate the flexural strength and interlaminar shear strength (ILSS) of fibrous composites. For this the ratio of the beam's length to thickness must be about 5 in order to suppress other forms of failure. The limitation of this test for the

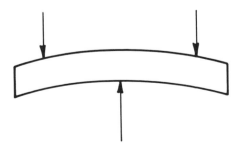

Figure 6.9. Beam in 3-point bending.

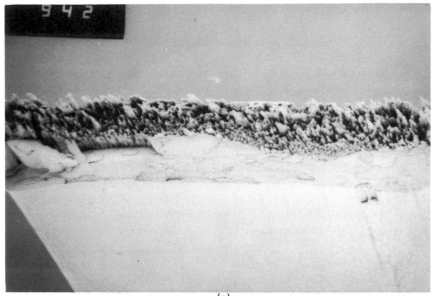

(a)

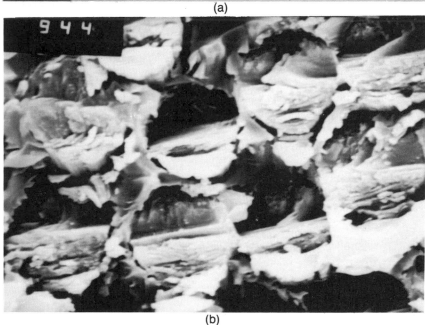

(b)

Figure 6.10. Scanning electron fractograph of a CFRP specimen tested in flexure at atmospheric pressure. (*a*) The fracture faces show approximately equal amounts of compressive and tensile type failure. (*b*) Higher magnification of a part of the compressive region showing demarcation of the fibre ends themselves into two regions, apparently tensile and compressive. (From Parry and Wronski, 1981—courtesy of Chapman and Hall Limited.)

measurement of ILSS has been recognized and alternative techniques have been proposed (see e.g. Markham and Dawson, 1975). Parry and Wronski (1981) have recently discussed in detail the failure processes occurring during flexural loading and the factors that influence them. They argue that the assumptions made in the derivation of equations (6.4) and (6.5) are not valid, and that this leads to an underestimate of the compressive stress carried by the beam. The situation is further complicated by the development of kink bands in the compressive half of the beam prior to failure. The kink bands are associated with the local high stresses developed at the point of application of the load to the specimen. The material behaves in a non-linear manner following the development of kink bands and, in the case of short thick beams, these eventually lead to the formation of interlaminar cracks. The interlaminar cracks are observed to follow resin-rich regions and do not necessarily propagate along the neutral plane of the beam.

Parry and Wronski also examined the effect of a superposed hydrostatic pressure on the failure in flexure of carbon-fibre–epoxy-resin composite beams having high and low length-to-thickness ratios and loaded in 3-point and 4-point bending. For the latter arrangement the stress distribution within the central portion of the specimen approximates to pure bending. In the case of short thick beams the formation of interlaminar cracks is suppressed at high hydrostatic pressures and compression kinking precedes failure. The same failure mode operates in the case of higher-aspect-ratio beams loaded in both 3 and 4-point bending. In all cases the boundary between the tensile and compressive portions of the fracture face moves towards the convex (compressive) face of the material as the superposed hydrostatic pressure is increased.

6.5 INTERLAMINAR SHEAR STRENGTH

Shear failure parallel with the fibres (in-plane shear strength, τ_{12}) occurs at a low stress level. As discussed in the previous section, it is often measured from the failure of a short, thick, unidirectionally reinforced beam loaded in flexure in 3-point bending, when the measured values and the details of the failure process depend strongly on the test conditions. Values measured using hoop-wound cylinders subjected to torsion and axial tension and compression (see §6.8) show the measured in-plane shear strengths to depend strongly on the magnitude of any transverse tensile or compressive stress applied at the same time.

Analytical studies of the stress distributions within idealized composites subjected to longitudinal shear forces indicate regions of high stress concentration occurring at the fibre–matrix interface at the points of least separation of the fibres (Adams and Doner, 1967a). These calculations suggest that very high shear stress concentrations will occur within localized regions of high fibre concentration, but these will be limited by local plastic or viscoelastic flow in the matrix. Experimentally, the shear strength is observed to depend on the strength of the interfacial bond and to be reduced by the presence of voids in the matrix.

6.6 OFF-AXIS LOADING OF A LAMINA

When a lamina is loaded in uniaxial tension at an angle θ to the fibre direction (Figure 6.11) various types of failure may occur depending on the value of θ. When θ is small the lamina would be expected to fail as a consequence of fibre fracture. When θ is near 90° transverse tensile failure would be expected. At intermediate values of θ failure would be expected to occur by in-plane shear.

The laminate strength as a function of θ can be computed very simply if it is assumed that failure will occur when the stresses or strains in one of the principal material directions reaches a critical value. The various modes of failure are assumed to operate independently of each other. The limitations of these assumptions are apparent, from the arguments outlined in the previous sections, and the predictions of failure stresses obtained are in only approximate agreement with experimental observations.

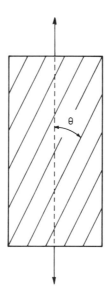

Figure 6.11. Off-axis tensile loading of a lamina.

The maximum stress theory postulates that failure will take place when any of the stresses in the principal material directions reaches a critical value. Thus, failure will occur when,

$\sigma_1 = X$ (load applied in the direction of fibre alignment);

$\sigma_2 = Y$ (load applied perpendicularly to the direction of
fibre alignment);

$\tau_{12} = S$ (shear failure parallel with the fibres).

Failure may take place in tension or compression and the numerical values of X and Y may be different for these different failure modes. Since the load may be applied at any angle θ to the direction of fibre alignment (see Figure 6.11), the stresses applied to the lamina must be transformed to stresses in the principal material directions. Both the component of the applied load and the cross-section within the lamina over which it is operating have to be computed. Thus if the applied stress is denoted by σ_θ the failure stress occurring by each of the three possible failure modes will be given by,

$$\sigma_\theta = X/\cos^2\theta$$

$$\sigma_\theta = Y/\sin^2\theta \qquad (6.6)$$

$$\sigma_\theta = S/\sin\theta\cos\theta$$

Since it is assumed that no interaction occurs between the three failure mechanisms, the composite strength is controlled by whichever operates at the lowest applied load. In Figure 6.12 a comparison is made between strength values predicted by the maximum stress theory and experimentally observed values of angle-ply silica-fibre-reinforced aluminium composites. The agreement between theory and experiment for this material is reasonable, particularly when allowance is made for the tendency of the fibres to rotate towards the direction of the applied load by matrix shear prior to composite failure. This effect is indicated by the arrows. Larger differences between the predictions of the maximum stress theory and observed off-axis strength values occur in the case of polymeric matrix composites. Observed strengths are appreciably less than the predicted values for a range of fibre orientation angles of around 30° when tensile loads are applied. This reflects the effects of combined transverse tension and in-plane shear stresses.

It can also be postulated that failure will occur when certain critical strain values are reached in the composite material. Again the various failure processes are assumed to operate independently of each other. The strains developed under off-axis loading are again resolved in the principal material directions and relationships similar to those developed for the maximum stress theory, but including a Poisson ratio term, can be produced (see Jones, 1975 for a discussion). The degree of correspondence between the

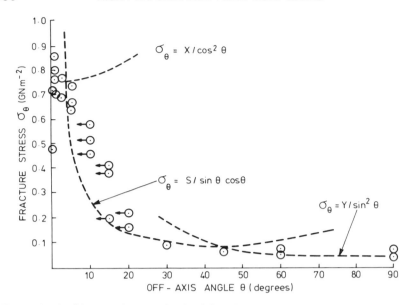

Figure 6.12. Observed strength of unidirectionally aligned aluminium–50% silica fibre composites as a function of fibre orientation compared to maximum stress theory. (Redrawn from Jackson and Cratchley, 1966.)

predictions of the maximum strain theory and experimental observations is comparable with that of the maximum stress theory.

Much better agreement between theory and experiment is obtained using an analysis developed by Tsai (1968) from a theory proposed by Hill (1950) to predict the yield stress of anisotropic metals. Hill assumed that the yield condition is a quadratic function of the stress components, so that

$$F(\sigma_y - \sigma_z)^2 + G(\sigma_z - \sigma_x)^2 + H(\sigma_x - \sigma_y)^2 + 2L\tau_{yz}^2 + 2M\tau_{zx}^2 + 2N\tau_{xy}^2 = 1 \tag{6.7}$$

where the state of anisotropy is characterized by the coefficients $F, G, H, M,$ and N and x, y and z are the axes of the assumed orthotropic material symmetry. These coefficients can be related to the failure strengths X, Y and S in the following way. If it is assumed that the anisotropic yield criteria can be equated with the anisotropic strength criteria of a laminate, equation (6.7) can be rewritten as

$$(G + H)\sigma_1^2 + (F + H)\sigma_2^2 + (F + G)\sigma_3^2 - 2H\sigma_1\sigma_2 - 2G\sigma_1\sigma_3 - 2F\sigma_2\sigma_3$$
$$+ 2L\tau_{23}^2 + 2M\tau_{13}^2 + 2N\tau_{12}^2 = 1 \tag{6.8}$$

We now consider that a stress σ_1 acts on the lamina, all other stresses being zero. The maximum value σ_1 can have is X, so that $(G + H) = 1/X^2$. From similar reasoning we see that $(F + H) = 1/Y^2$ and $(F + G) = 1/Z^2$, where Z is the strength in the 3 direction. If only τ_{12} acts on the body, then $2N = 1/S^2$. From considerations of geometric similarity, $Y = Z$. If we consider plane stress conditions only, equation (6.8) can now be written as

$$\frac{\sigma_1^2}{X^2} - \frac{\sigma_1\sigma_2}{X^2} + \frac{\sigma_2^2}{Y^2} + \frac{\tau_{12}}{S^2} = 1 \qquad (6.9)$$

which is known as the Tsai–Hill failure criterion.

If it is assumed that failure under compressive stresses can be treated in the same way, equation (6.9) can be used to describe the failure envelope in the four quadrants of the $\sigma_1-\sigma_2$ stress space. X^1 and Y^1 are substituted for X and Y when compressive stresses occur. For unidirectional off-axis loading the stress transformation equations (6.6) are applied, so that equation (6.9) becomes

$$\frac{\cos^4\theta}{X^2} + \left(\frac{1}{S^2} - \frac{1}{X^2}\right)\cos^2\theta \sin^2\theta + \frac{\sin^4\theta}{Y^2} = \frac{1}{\sigma_\theta^2} \qquad (6.10)$$

Equation (6.10) has been shown to predict quite accurately the off-axis tensile and compressive strengths of E glass epoxy composites (Tsai, 1968). In Figure 6.13 comparison is made between the predictions of equations (6.6) and (6.10) and experimental data.

The approach used to derive equation (6.9) leads to a condition in which only the longitudinal strength X of the lamina is involved in the term dealing with the interactions between stresses in the 1 and 2 directions. A more general theory has been derived by Tsai and Wu (1971) and this is discussed by Tsai and Hahn (1980). The basic assumption of this failure criterion is that a failure surface in stress space exists such that

$$F_i\sigma_i + F_{ij}\sigma_{ij} = 1 \qquad (i, j = 1, 2 \ldots 6) \qquad (6.11)$$

where F_i and F_{ij} are strength tensors of the second and fourth rank respectively. When applied to an orthotropic lamina subjected to plane stress, equation (6.11) can be written

$$F_1\sigma_1 + F_2\sigma_2 + F_6\sigma_6 + F_{11}\sigma_1^2 + F_{22}\sigma_2^2 + F_{66}\sigma_6^2 + 2F_{12}\sigma_1\sigma_2 = 1 \quad (6.12)$$

where $\sigma_6 = \tau_{12}$.

Some of the components of the strength tensors can be defined in terms of the engineering strengths of the material. Since we assume the shear strength in principal material directions is independent of the sign of the shear stress σ_6 we have, setting other stresses to zero, $F_6 = 0$ and $F_{66} = 1/S^2$, where S is the shear strength of the material. Similarly, at

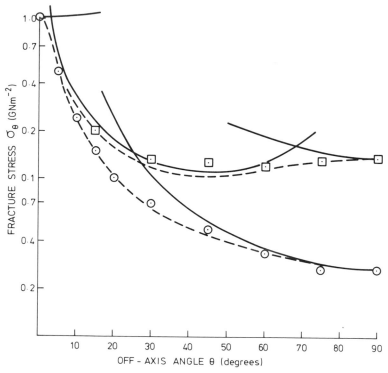

Figure 6.13. Experimental strength values for glass-fibre–epoxy-resin unidirectional lamina tested in uniaxial tension (◯) and compression (□). Full curves calculated from equation (6.6); broken curves from equation (6.10). (Redrawn from Tsai, 1968.)

failure under a uniaxial tensile load X_t applied in the 1 direction, we have

$$F_1 X_t + F_{11} X_t^2 = 1$$

and for failure in uniaxial compression we have

$$F_1 X_c + F_{11} X_c^2 = 1$$

so that

$$F_1 = \frac{1}{X_t} + \frac{1}{X_c} \quad \text{and} \quad F_{11} = \frac{-1}{X_t X_c}$$

Similarly,

$$F_2 = \frac{1}{Y_t} + \frac{1}{Y_c}$$

and

$$F_{22} = \frac{-1}{Y_t Y_c}$$

The value of the fourth-rank tensor term F_{12} has to be established from tests carried out under biaxial loading, and this presents greater experimental problems than those encountered in the measurement of the other terms. Pipes and Cole (1973) found the predictions of equation (6.12) not to be sensitive to the numerical values assigned to F_{12}. However, in order that equation (6.12) should represent a closed surface under all combinations of stress (see §6.8), $F_{12}^{\star}$ given by

$$F_{12}^{\star} = F_{12}/(F_{11}F_{22})^{1/2} \qquad (6.13)$$

must lie between $+1$ and -1. Tsai and Hahn (1980) point out that if $F_{12}^{\star} = -\frac{1}{2}$, equation (6.12) is consistent with the von Mises criterion for isotropic materials.

The advantages of the Tsai–Wu tensor failure theory (equation (6.12)) over the Tsai–Hill failure criterion (equation (6.9)) stem from its more general character and similarity to the stiffness and compliance tensors. This allows mathematical operations such as the transformation of axes to be carried out using known tensor transformation laws. However, it should be borne in mind that such failure theories do not take into account the physical processes associated with failure. These change according to the combinations of applied stresses and also depend upon the details of the microstructure of the composite. The applicability of general theories of fracture is therefore less certain than that of similar theories dealing with the elastic properties of composites.

6.7 STRENGTH OF LAMINATES

In the elementary theory of laminates outlined in §3.8 it is assumed that each ply is perfectly bonded to its neighbour and that all experience the same strain values within the plane of the laminate. The stress levels developed in any particular lamina can be calculated and these, in general will be different since they depend upon the orientation of the fibres in the individual plies. If the stresses developed in individual laminae are compared with a particular failure criterion (for example, equation (6.6)), the overall stress carried by the laminate at which one or more individual plies fail can be calculated.

Interactions between the laminae may suppress or initiate the failure of individual plies. In the case of $0°/90°$ glass-fibre–epoxy-resin laminates, for example, a load applied in the $0°$ direction can cause the $0°$ plies to split parallel to the fibres and parallel to the direction of loading (Bailey *et al.*, 1979). This mode of failure is due to a combination of elastic and thermal effects and would not occur if the $0°$ plies were loaded independently. Interaction effects between adjacent plies can also enhance the strain values at which the cracking of individual plies occurs through modifications to the mechanics of crack growth. This effect depends on the absolute thickness of the individual plies and becomes apparent only at small ply thicknesses.

Initially we assume that any particular lamina within a laminate will fail when the stresses which it is supporting reach a value at which one or other of the possible failure modes discussed in §6.6 operates. When one lamina fails, additional loads are applied to the still-intact laminae. This process continues until the remaining intact laminae are no longer able to support the applied load. The overall elastic behaviour of the laminate will be non-linear if several laminae fail sequentially before eventual gross fracture occurs.

The procedure to be followed in the analysis of laminate fracture is given in outline below with particular reference to angle-ply and cross-ply systems. More comprehensive treatments can be found in appropriate text books on the theory of laminates (e.g. Jones, 1975; Ashton *et al.*, 1969). For simplicity we consider a laminate having a mid-plane symmetry subjected to a tensile load. The A matrix is first constructed for the laminate as described in §3.8. This matrix gives the stress strain equation of the laminate and is inverted to give the strain–stress relationships, so that when $N_x = N_1$ and $N_y = N_{xy} = 0$, we have,

$$\begin{Bmatrix} \varepsilon_x^0 \\ \varepsilon_y^0 \\ \gamma_{xy}^0 \end{Bmatrix} = \begin{bmatrix} A_{11}' & A_{12}' & A_{16}' \\ A_{12}' & A_{22}' & A_{26}' \\ A_{16}' & A_{26}' & A_{66}' \end{bmatrix} \begin{Bmatrix} N_1 \\ 0 \\ 0 \end{Bmatrix}$$

where the A' matrix is the inverse of the A matrix given in equation (3.34), so that

$$\begin{Bmatrix} \varepsilon_x^0 \\ \varepsilon_y^0 \\ \gamma_{xy}^0 \end{Bmatrix} = \begin{Bmatrix} A_{11}' \, N_1 \\ A_{12}' \, N_1 \\ A_{16}' \, N_1 \end{Bmatrix}$$

The stresses developed in each ply are then obtained from the stress–strain relationship for a lamina (equation (3.16)):

$$\begin{Bmatrix} \sigma_x \\ \sigma_y \\ \tau_{xy} \end{Bmatrix} = \begin{bmatrix} \bar{Q}_{11} & \bar{Q}_{12} & \bar{Q}_{16} \\ \bar{Q}_{12} & \bar{Q}_{22} & \bar{Q}_{26} \\ \bar{Q}_{16} & \bar{Q}_{26} & \bar{Q}_{66} \end{bmatrix} \begin{Bmatrix} A_{11}' \, N_1 \\ A_{12}' \, N_1 \\ A_{16}' \, N_1 \end{Bmatrix}$$

The stresses and corresponding strains in the x, y directions can now be calculated for each lamina and can be compared with the failure criteria previously discussed.

In the case of angle-ply laminates the failure stress for the laminate is given by the first ply-failing stress, since all of the plies fail simultaneously in this class of laminate. The Tsai–Hill criterion (equation (6.9)) has been shown to predict quite accurately the failing stresses of angle-ply laminates (Tsai, 1968). The strength shows a similar dependence on the ply orientation $\pm\theta$ as the unidirectional lamina strength (see Figure 6.13) but is rather larger for values of θ between 0° and 45° than is the case for a single ply.

The stresses developed in cross-ply symmetric laminates, loaded in the principal material directions, can be computed quite straightforwardly because the $\bar{Q}_{ij}$ values given in equations (3.17)–(3.22) take a very simple form for this condition. Thus

$$
\left.
\begin{aligned}
\bar{Q}_{11} &= Q_{11} \\
\bar{Q}_{12} &= Q_{12} \\
\bar{Q}_{22} &= Q_{22} \\
\bar{Q}_{66} &= Q_{66} \\
\bar{Q}_{16} &= 0 \\
\bar{Q}_{26} &= 0
\end{aligned}
\right\}
\text{(for the longitudinal ply) and}
\left\{
\begin{aligned}
& Q_{22} \\
& Q_{12} \\
& Q_{11} \\
& Q_{66} \\
& 0 \\
& 0
\end{aligned}
\right.
\text{(for the transverse ply)}
$$

If we assume that the longitudinal and transverse ply thicknesses are the same ($V_A = V_B = 0.5$), the $\bar{Q}_{Lij}$ terms in equation (3.37) can be written

$$
\bar{Q}_{L11} = (Q_{11} + Q_{22})/2, \quad \bar{Q}_{L12} = Q_{12}
$$
$$
\bar{Q}_{L22} = (Q_{22} + Q_{11})/2, \quad \bar{Q}_{L66} = Q_{66}
$$

The Q_{ij} terms can be calculated from the engineering elastic properties of the individual laminae using equations (3.12) and the engineering elastic properties of the cross-ply laminate can be calculated from the $\bar{Q}_{Lij}$ terms using the same relationships. In Table 6.2 the engineering elastic properties of a single lamina and the engineering elastic properties of a symmetric cross-ply system containing equal volume fractions of such laminae are shown. The properties of the laminae are the same as those used in computing the elastic properties of $\pm\theta$ laminates shown in Figure 3.14.

We now consider the stresses developed in two adjacent plies of equal thickness forming part of a symmetric cross-ply laminate. If a stress is applied in one of the principal material directions, the laminate will extend in the direction of the applied load and contract laterally under the influence of the laminate Poisson ratio. The same extensions could be developed in

TABLE 6.2
Engineering elastic properties of laminae and cross-ply laminate

Lamina	Laminate
$E_1 = 241 \text{ GN m}^{-2}$	$E_1 = 124.6 \text{ GN m}^{-2}$
$E_2 = 7.72 \text{ GN m}^{-2}$	$E_2 = 124.6 \text{ GN m}^{-2}$
$G_{12} = 6.13 \text{ GN m}^{-2}$	$G_{12} = 6.13 \text{ GN m}^{-2}$
$v_{12} = 0.27$	$v_{12} = v_{21} = 0.0168$
$v_{21} = 0.00865$	

the individual laminae if they were subjected to suitable external loads applied in the principal material directions. These loads and corresponding stresses can be calculated quite straightforwardly. They correspond to the in-plane stresses developed in the individual laminae of the bonded cross-ply laminate.

If a tensile strain ε is developed in the direction of the applied load, the average stress developed in the laminate in the direction of loading will be $124.6\varepsilon \text{ GN m}^{-2}$. The stress carried by the longitudinal and transverse plies in this direction will be $241\varepsilon \text{ GN m}^{-2}$ and $7.72\varepsilon \text{ GN m}^{-2}$ respectively. Under load the laminate will contract laterally by 0.0168ε, so that this strain will be developed in a direction perpendicular to the fibre orientation in the longitudinal ply (see Figure 6.14). The Poisson ratio for the longitudinal ply is 0.27 so that the contraction strain which would have been developed in the longitudinal ply in the transverse direction if unbonded and not under constraint would be 0.27ε. The strain difference of 0.253ε corresponds to a stress in the longitudinal ply of $7.72 \times 0.253\varepsilon \text{ GN m}^{-2}$ or $1.96\varepsilon \text{ GN m}^{-2}$, which tends to split the longitudinal ply in directions parallel to the fibre direction. The stresses developed in the transverse ply can be computed in a similar manner, since a compressive strain given by $(0.00865 - 0.0168)\varepsilon$ is imposed on the transverse ply in the fibre direction. Thus a compressive stress of $1.96\varepsilon \text{ GN m}^{-2}$ is developed parallel with the fibres in this ply balancing the corresponding tensile stress developed in the other ply.

A resin matrix composite is normally fabricated at temperatures above those at which it will be used. Owing to differential thermal contraction, stresses are developed in addition to those due to elastic anisotropy discussed above. Their magnitudes are influenced by any relaxation that may take place in the polymer during curing and subsequent cooling and can be measured conveniently using a 2-ply 0°/90° test specimen. This is flat as fabricated but becomes curved, rather like a bimetallic strip, on cooling to room temperature. The stresses developed by differential thermal contraction can be measured from such a specimen using simple beam

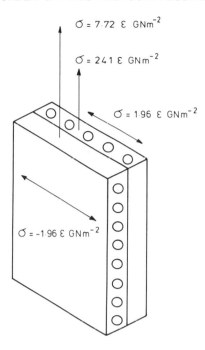

Figure 6.14. Stresses developed in a cross-ply laminate due to a tensile load applied parallel to the alignment of the fibres in one of the lamina.

theory (Bailey *et al.*, 1979). The curvature of the strip decreases with increasing temperature and in this way the effective curing temperature of the laminate can be estimated (Flaggs and Kural, 1982).

6.8 FAILURE OF LAMINATES UNDER COMBINED IN-PLANE STRESSES

In the preceding sections the failure of a laminate under unidirectional loading has been considered. It has been pointed out that the failure of a lamina may occur by fibre fracture, by in-plane tensile failure of the matrix (or fibre–matrix interface), or by the in-plane shear failure of the matrix (or fibre–matrix interface). Depending on the orientation of a lamina, any of these types of failure may be induced by the application of a tensile load.

When used in engineering structures, fibre-reinforced laminates are usually subjected to a combination of stresses. It is experimentally more difficult and time-consuming to carry out strength tests under biaxial loading conditions and, hence, experimental data relating to combined stresses are relatively scarce. Investigations have been confined to plane stress conditions which are representative of most applications of laminates, since these are generally used in sheet form. Most of the investigations carried out have utilized thin-walled tubes subjected to internal hydrostatic pressure plus axial and sometimes torsional loads. Specimens can be manufactured conveniently by filament winding onto a mandrel and woven fabric reinforcement can be used to fabricate tubular specimens. When the tube is subjected to hydrostatic pressure an internal membrane is required to prevent leakage of the pressurizing fluid through the tube wall, which usually becomes porous prior to failure. Owen and co-workers (see e.g. Owen and Rice, 1981b), have used a system in which oil under pressure is applied both to the tube and to a loading ram that can be arranged to generate axial tensile or compressive loads. Various rams are used to generate different stress ratios. Hull et al. (1978) have applied internal pressures to tubes (a) with their ends restrained so that the axial strain of the tube was zero, or (b) with their ends sealed so that the hoop stress in the tube was twice the axial stress, or (c) with their ends free to slide within hydraulic seals so that the axial stress is always zero. Other workers have applied axial tension and torsion to cylinders (Puck & Schneider, 1969).

To predict safe design loads for composite structures under complex stress conditions, a failure theory is required. This defines a failure envelope in stress space such that failure will not occur for all combinations of stress that lie within the envelope. Many such relationships between strength and combined stresses have been proposed. It is important to realize that these failure theories do not take into account details of the mechanisms of failure of the various failure modes, which will change as the ratio between the various stresses changes. In §6.6 we discussed the application of failure criteria defined by equation (6.9) to the off-axis tensile and compressive loading of a lamina. If a hoop-wound cylinder is subjected to torsion and an axial tensile force, then σ_1 is zero, so that equation (6.9) reduces to

$$\left(\frac{\sigma_2}{Y}\right)^2 + \left(\frac{\tau_{12}}{S}\right)^2 = 1 \qquad (6.14)$$

which is the equation of an ellipse. When experimental values for the longitudinal shear strength S and transverse tensile strength Y are inserted, equation (6.14) is found to predict the strength of a laminate under combined shear and transverse tensile stresses with reasonable accuracy. When axial compression loading is included, the data can be represented by an equation of the form

$$-\frac{\sigma_2^2}{Y_t Y_c} + \sigma_2 \frac{(Y_c + Y_t)}{Y_t Y_c} + \left(\frac{\tau_{12}}{S}\right)^2 = 1 \qquad (6.15)$$

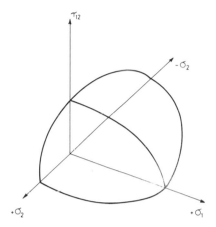

Figure 6.15. Portion of failure surface in σ_1, σ_2 and τ_{12} space.

where Y_t and Y_c are the failure strengths in transverse tension and transverse compression. The sign convention used is that the stresses are written positively in the algebra and assume negative numerical values in compression. Greenwood (1977) has discussed the validity of equation (6.15) and other more complex expressions.

When the effects of additional loads applied in a direction parallel to the fibre alignment are included, fracture conditions are defined by a 3-dimensional failure surface. Fracture will not occur for combinations of stress represented by points within the failure surface; Equation (6.12) defines such a surface, part of which is shown schematically in Figure 6.15. This should be a closed surface under all combinations of stress, otherwise stress conditions giving infinite strengths would be implied. Also, the origin of the coordinate system should lie within the surface, or fracture would be implied at zero stress. A large number of relationships giving failure surfaces for combined stresses have been proposed. Tsai and Hahn (1980) suggest that they can be encompassed within a general quadratic function of the form given in equation (6.12). This equation can be rewritten in the form

$$\left(\frac{1}{X_t}-\frac{1}{X_c}\right)\sigma_1 + \left(\frac{1}{Y_t}-\frac{1}{Y_c}\right)\sigma_2 + \frac{\sigma_1^2}{X_tX_c} + \frac{\sigma_2^2}{Y_tY_c} + 2F_{12}\sigma_1\sigma_2 + \frac{\tau^2}{S^2} = 1 \quad (6.16)$$

As discussed in §6.6, it is necessary for $F_{12}/(F_{11}F_{22})^{1/2}$ to lie between $+1$ and -1 in order that equation (6.12) should form a closed surface and that it is convenient to assign to this ratio a value of $-\frac{1}{2}$. A considerable amount of data is required in order to assess the experimental validity of a suggested failure surface and a large number of mathematical relationships have been proposed. Owen and Rice (1981) have compared many of these with experimental data obtained from polyester resins reinforced with glass fibres in the form of woven fabric. It is important to appreciate that existing failure theories do not take account of the details of the fracture mechanisms, which will be different at different points on the fracture surface and will also be influenced by details of the composite microstructure.

One very simple approach to the design of laminated structures is to assume that the fibres support all of the applied load, the contribution of the matrix being negligible (netting analysis). This approach can be applied to filament-wound tubes. If a closed-end tube is considered, the axial stress σ_A in the tube wall will be one-half of the hoop stress σ_H. The relative contributions of the hoop stress and the axial stress to the stress carried by the reinforcing fibres will vary with the filament winding angle. From the transformation equations (3.15), after neglecting any shear and transverse tensile contributions, we see that mutually perpendicular stresses σ_x and σ_y produce a stress σ in the fibres, where

$$\sigma = \sigma_x \cos^2\theta + \sigma_y \sin^2\theta$$

and θ is the angle between the fibres and the x axis. If σ_x corresponds with the tube axis and σ_y with the hoop direction, we have $\sigma_y = 2\sigma_x$ for a closed tube under internal pressure. The optimum value of θ is obtained when $\sin^2\theta = 2\cos^2\theta$ and is therefore given by $\tan^2\theta = 2$ or $\theta = 55°$. The filament winding process of course produces fibre orientations of $\pm\theta$.

The elastic modulus of a filament-wound tube is observed to be much higher than that predicted by the netting analysis. This of course is due to the contribution of the matrix. Cracks are formed in the matrix at low strain values and these increase in number as the internal pressure is increased. Each crack initially develops within one individual lamina, and they first occur in local regions of high fibre content (Jones and Hull, 1979). This is an expected consequence of the presence of high stress concentrations at the fibre–matrix interface and within the matrix between closely spaced fibres. The cracks do not propagate into adjacent laminae but, when sufficient cracks are present, liquid can escape from the pipe through paths formed by the intersection of transverse cracks in adjacent laminae. The number of cracks increases with increasing hydrostatic pressure; this is consistent with the theories of transverse cracking in laminates. Interlaminar cracking also occurs in filament-wound pipes and becomes more pronounced as the pressure is increased; this process enhances the leakage of liquid from the pipe. As cracking proceeds the contribution of the matrix to the properties of the laminate falls and the measured values of elastic modulus approach those predicted by the netting analysis. Catastrophic failure involving extensive fibre fracture eventually occurs (Hull *et al.*, 1978).

6.9 EDGE EFFECTS IN LAMINATES

In the classical theory of laminates introduced in Chapter 3, only stresses in the plane of the laminae (σ_x, σ_y and τ_{xy}) were considered, and the out-of-plane stresses σ_z, τ_{zx} and τ_{zy} were ignored. This simplification is not justified at positions near to the free edge of a laminate where large out-of-plane stresses can be developed (see e.g., Pipes and Pagano, 1970). These stresses are localized and fall to zero at distances from the edge approximately equal to the laminate thickness. The magnitudes of the out-of-plane stresses depend upon the construction of the laminate in terms of the proportion, thickness, fibre orientation and stacking sequence of laminae (Pagano and Pipes, 1971). Curtis (1984) argues that the stress σ_z acting in a direction perpendicular to the plane of the laminae is the main factor controlling edge cracking in 0°/45°/90° laminates. Laminates can be designed such that, when loaded in tension in the plane of the laminate, the values of σ_z developed at the laminate edge are compressive. The stress σ_z then becomes tensile when compressive loads are applied, so that for most applications a stacking sequence that minimizes the numerical value of σ_z is required. The residual stresses developed by cooling the laminate from its fabrication temperature and those produced by the absorption of moisture by an epoxy resin matrix are also significant in the development of out-of-plane stresses and associated edge cracking.

7
The Long-term Mechanical Properties of Composites in Engineering Applications

The mechanical properties of a structural composite are degraded under service conditions. The strength may be reduced owing to localized stress enhancement as a consequence of mechanical damage. Strengths and elastic moduli are generally reduced by fatigue damage resulting from the repeated application of a load that is initially insufficient to cause failure. Damage resulting from environmental effects can also occur. A comprehensive treatment of these issues is beyond the scope of this book. In this chapter a few examples of these effects are discussed in outline and the references cited provide an introduction to the extensive technical literature on these subjects.

Composites are widely used in the form of laminates and it is important to be able to predict their residual strength after they have suffered mechanical damage. The development of fracture mechanics has led to an understanding of the damage tolerance of metal structures. The principles involved are outlined in §7.1. Fracture mechanics has proved to be less useful in predicting in detail the strength of damaged laminates because of the much greater complexity of the failure process. It is, however, widely used to provide a numerical basis of comparison and with various modifications to predict the strength of damaged laminates. This topic is discussed in §7.2.

Fatigue failure is discussed in §7.3; §7.3.1 deals with unidirectional systems and §7.3.2 with laminates. Some aspects of the environmental degradation of composites are outlined in §7.4.

7.1 FRACTURE MECHANICS

Brief reference was made in §1.2 to the enhanced stresses developed at the tip of a notch or crack in an isotropic elastic material under load. These local stresses increase as the crack or notch length increases and as the radius of curvature of its tip decreases, so that for cracks of appropriate geometry they may be large enough to cause a brittle material to fracture or an elastic/plastic material to deform at the notch tip. According to the Griffith concept, also outlined in §1.2, an existing crack will propagate if thereby the total energy of the system is lowered. Energy is supplied by the elastic relaxation of the system during crack extension. Energy is absorbed in the creation of fracture surfaces and, particularly in the case of metals, by the plastic deformation of the material in the vicinity of the crack tip prior to fracture.

The concept of local stress concentration implies that as the radius of curvature of the tip of a notch is reduced to interatomic distances, i.e. as the notch becomes a crack, the local stress approaches infinity (equation (1.7)). Thus it is difficult to apply this concept to the prediction of the stress that causes failure of a material containing a crack of arbitrary length. On the other hand, the rate of release of strain energy with increasing crack length can be obtained by direct experiment (see for example Paris and Sih, 1965). Under load the strain energy U in an elastic body containing a crack of arbitrary length a (see Figure 1.4(a)) is given by

$$U = Q^2 c/2 \qquad (7.1)$$

where Q is the applied force, per unit thickness and c the compliance of the body defined by $\varepsilon = cQ$, where ε is the extension of the body under the action of the force Q. Hence the rate of release of strain energy with respect to crack extension is given by

$$\frac{\partial U}{\partial a} = \tfrac{1}{2} Q^2 \frac{\partial c}{\partial a} \qquad (7.2)$$

The compliance of a test specimen, or a model of a component, can be obtained experimentally over a range of crack lengths. Thus the value of $\partial c/\partial a$ can be deduced as a function of the crack length. From a knowledge of the load needed to cause failure of a specimen containing a crack and the value of $\partial c/\partial a$ at the appropriate crack length, $\partial U/\partial a$ at fracture can be evaluated. This critical value of the rate of release of strain energy can be equated with G_{1c} (equation (1.11)) for plane stress conditions. The compliance of various shapes containing cracks of arbitrary length can be calculated analytically, enabling predictions to be made of the load-bearing ability of a cracked component. Later analytical developments, based on the

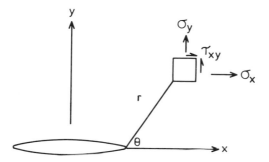

Figure 7.1. Coordinates measured from the leading edge of a crack and stress components of the crack tip stress field.

stress intensity concept, are more convenient to apply and have replaced the compliance method for most applications. These are discussed in outline below.

By making use of analyses previously developed by Westergaard, Irwin (1957) noted that the stresses in the vicinity of a crack tip, as illustrated in Figure 7.1, could be expressed in the following way:

$$\sigma_x = \frac{K_1}{(2\pi r)^{1/2}} \cos (\theta/2)(1 - \sin (\theta/2) \sin (3\theta/2))$$

$$\sigma_y = \frac{K_1}{(2\pi r)^{1/2}} \cos (\theta/2)(1 + \sin (\theta/2) \sin (3\theta/2)) \qquad (7.3)$$

$$\tau_{xy} = \frac{K_1}{(2\pi r)^{1/2}} \cos (\theta/2)(\sin (\theta/2) \cos (3\theta/2))$$

where r and θ are the cylindrical polar coordinates of a point with respect to the crack tip. This analysis refers to Mode I (crack opening) conditions. Similar relationships can be derived for crack propagation under Mode II and Mode III conditions, which involve the parameters K_{11} and K_{111} (see e.g. Paris and Sih, 1965). These equations are derived by ignoring higher-order terms in r and can thus be regarded as being approximately correct in the regions where r is small compared with the crack length and other dimensions of the body.

The parameters K_1, K_{11} and K_{111} are termed stress intensity factors. They are not dependent on the coordinates r and θ and hence they control the intensity of the stress fields but not their distribution for each mode. For an infinite plate subjected to a uniform tensile stress and containing a transverse crack of length $2a$ (Figure 7.1), it can be shown that K_1 is given by

$$K_1 = \sigma(\pi a)^{1/2} \tag{7.4}$$

The stress intensity factor K is a function of the applied stress and crack geometry and the crack propagates when K reaches a critical value K_c. The equivalence of G_{1c} and K_{1c} was derived by Irwin (1957), and for plane stress conditions is given by $K_{1c}^2 = EG_{1c}$, where E is the Young's modulus of the material. Similar relationships are found for other modes of crack opening, crack geometries and loading conditions, and for orthotropic and anisotropic materials (see e.g. Paris and Sih, 1965). In general stress intensity factors may be written in the form:

$$K_1 = \sigma a^{1/2} Y(a/W) \tag{7.5}$$

where a is a characterizing crack length, σ is a characterizing stress, W is a characterizing dimension and $Y(a/W)$ is a calibration function for the specific body under consideration. For example, if the plate shown in Figure 7.1 has a finite width $2b$ an approximate value for K_1 is given by

$$K_1 = \sigma(\pi a)^{1/2} \left[\frac{2b}{\pi a} \tan \frac{\pi a}{2b} \right]^{1/2} \tag{7.6}$$

The term within the square brackets approaches unity when b becomes much greater than a. In the case of an edge crack in a semi-infinite plate, the value of the stress intensity factor K_1 is given approximately by

$$K_1 = 1.12 \, \sigma(\pi a)^{1/2} \tag{7.7}$$

The stress intensity factor approach to the prediction of the conditions for fracture is based on the assumption that the mechanical environment in the immediate vicinity of the crack tip controls the fracture behaviour. Only a limited amount of plastic deformation is assumed to occur, so that this approach to the analysis of fracture is termed linear elastic fracture mechanics (LEFM). The geometry and loading conditions influence the conditions at the crack tip through the parameter K, which is determined by a suitable analysis. Fracture occurs at a critical value of the stress intensity factor K_c. The value of K_c can be obtained using laboratory test specimens and the data can be used to predict the fracture stresses for other bodies containing cracks for which the K values are the same but which may be generated by different combinations of load and crack length.

In the case of metals, a plastic zone is generated ahead of the geometrical crack tip. The size of this zone can be estimated from the value of r obtained from equation (7.3) after setting σ_x equal to σ_{ys}, the yield strength of the material. Providing the size of this zone is negligible compared with the volume of material undergoing elastic deformation, the fracture process is described adequately by LEFM and the total effective crack length is obtained by adding the size of the plastic zone to the length of the geometrical crack.

Consideration has also to be given to the effect of sample thickness. In the case of thick plates (plane strain conditions) the size of the plastic zone is reduced away from the free surfaces because of the constraint exerted by the surrounding material. Measured values of the critical rate of strain energy release and the critical stress intensity factor, decrease with increasing sample thickness, eventually reaching a limiting value as plane strain conditions are approached.

It is found that the fracture toughness of a metal is best characterized using thick specimens, for which the crack growth takes place essentially under conditions of plane strain. The development of linear elastic fracture mechanics, as outlined above, has made it possible to predict with reasonable precision the residual strength of damaged metal structures and the concept of the stress intensity parameter K has been found to be particularly useful when extensive plastic deformation does not occur.

7.2 NOTCH SENSITIVITY OF LAMINATES

The inhomogeneous and anisotropic characteristics of fibre composites generate a wide range of failure processes. These depend on the properties of the fibres, matrix and interface and also on the orientation thicknesses and stacking sequence of the plies in a laminate. It follows that the analysis of the notched or damaged strength of laminates is more complex than that of metals. The nature of the problem can be seen from the following simple examples.

We first consider a single unidirectionally reinforced polymeric matrix lamina containing a notch aligned parallel to the direction of the fibres. When a crack-opening tensile force is applied at right-angles to the fibre alignment, the crack propagates in a self-similar manner parallel to the fibres. This situation can be analysed using LEFM. However, if the notch is now arranged to be perpendicular to the fibres, so that the crack-opening force is applied in the fibre direction, shear cracks initiate at the ends of the notch and propagate parallel to the fibres. The lengths of the shear cracks increase as the applied stress increases; the overall effect is to remove the local stress concentration developed by the initial notch (Figure 7.2).

When additional off-axis plies are added to the 0° ply, illustrated in Figure 7.2, their effect is to suppress the growth of the shear cracks. The laminate now becomes notch-sensitive because of the increased notch sensitivity of the 0° ply, which supports most of the applied tensile load. The magnitude of this effect is dependent on the relative thickness of the

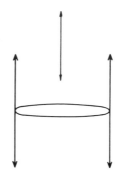

DIRECTION OF FIBRE ALIGNMENT
AND APPLIED LOAD

Figure 7.2. Illustrating the development of shear cracks in a notched unidirectionally reinforced fibre composite.

plies, the ply stacking sequence and the degree of bonding between the fibres and the matrix. The effect of ply thickness is particularly apparent; the notch sensitivity of the 0° plies of a laminate diminishing and the laminate fracture toughness thus increasing as the thickness of the individual plies forming the laminate are increased.

At some threshold stress level, cracks are initiated at the tip of a notch in a laminate. These propagate parallel to the fibres in the various plies as the load is increased. In this way a zone of damage is produced, the shape of which depends on the structure of the laminate. A sketch of the damage zone developed in an edge-notched glass-fibre–epoxy-resin laminate is shown in Figure 7.3. The size of the damage zone, measured by the lengths of the subcracks in the various plies, increases as the applied stress increases and was found to be approximately proportional to calculated values of K_1^2 by Mandell *et al.* (1975). The fracture toughness, measured as the stress intensity value at fracture and calculated on the basis of the initial notch length, is also observed to increase as the size of the damage zone increases. These parameters are also observed to increase as the length of the initial notch is increased. The lengths of the subcracks developed in the various plies within the damage zone are different for different ply orientations and laminate constructions (Mandell *et al.* 1975; Ochiai and Peters, 1982a; Peters, 1983). Although the elastic anisotropy generally present in laminates can be accommodated in the LEFM analysis, the propagation of the subcracks in the individual plies occurs under a complex combination of failure modes. Furthermore, the degree of interaction between adjacent plies is also influenced by the amount of delamination occurring within the damage zone. In many cases the size of the damage zone is comparable with

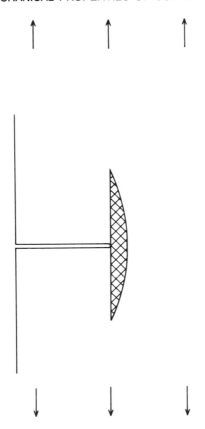

Figure 7.3. Sketch illustrating the development of a typical damage zone near the tip of an edge crack in a cross-ply laminate subjected to a tensile load.

the length of the initial notch and in the case of ±45° laminates it can spread across the entire width of a test panel before complete rupture occurs (Mandell *et al.* 1975). Although the lengths of the subcracks within the damage zone are observed to increase with increasing values of K (calculated on the basis of the initial notch length) the precise relationship between these quantities is not fully established (Ochiai and Peters, 1982b; Peters, 1983). It is generally observed that the stress intensity value at fracture (fracture toughness) increases as the thickness of the individual plies increase, but it is not sensitive to the overall thickness of the laminate. These phenomena are consistent with the supposition that the strength of the laminate is controlled primarily by the 0° plies and that the notch sensitivity of these plies is enhanced by the constraints exerted by adjacent angle plies and reduced by any delamination effects occurring at the inter-ply interfaces. When a laminate is subjected to in-plane stresses, shear stresses and normal stresses are developed at the inter-ply interfaces (Pipes and Pagano, 1970). These are enhanced at the tip of a notch, where the in-plane stresses are greatest, and also depend on the ply stacking arrangements and ply thicknesses. The interlaminar shear stresses and tensile stresses tend to produce delamination effects, thus reducing the notch sensitivity of the 0° primary load-bearing plies. It is found that LEFM can be used to calculate the strength of notched composites from the critical stress intensity factor at failure when the size of the damage zone is small, but LEFM is less useful when damage zones of appreciable size are developed. In some circumstances the value of the stress intensity factor at fracture is observed to increase with increasing crack length (Peters, 1983).

The details of the processes by which notches reduce the strength of laminates is not fully understood at present. One effect that has received considerable attention is the reduction in the strength of a laminate due to the presence of a circular hole. It is found that, over a range of diameters up to about 30 mm, the strength of a laminate falls with increasing size of the hole. This behaviour cannot be explained in terms of the elastic stress concentration factor, which is numerically equal to 3 whatever the size of the hole. When the hole is large the strength is reduced to about one-third of the nominally undamaged value so that, for these conditions, the behaviour approximates to that predicted for an elastic material. A number of analytical approaches have been developed to account for this behaviour. If it is assumed that cracks are developed in the highly stressed regions adjacent to the hole, LEFM can be used to predict the strength of the laminate. The calculation is based on the stress intensity factor associated with cracks that have extended by some characteristic distance a from the edge of a hole (Waddoups et al., 1971) (Figure 7.4). Alternatively, it has been proposed that failure will occur when the average stress developed over some characteristic distance a_0 from the edge of the hole, or reached at a point some characteristic distance d_0 from the edge of the hole, equals the unnotched strength of the material (Whitney and Nuismer, 1974). In these approaches it is assumed that the characteristic distance a associated with the first model and a_0 and d_0 associated with the second model can be treated as material constants. The analysis has also been used to predict the strength reduction due to narrow slits. When suitable values for a, a_0 or d_0 are chosen (usually of the order of 1 mm), these analyses have been shown to predict the strength of damaged laminates of various types with reasonable accuracy. However, because of the complex nature of the failure processes occurring in laminates, this type of model would not be expected to be universally applicable and it has indeed been observed to be unsatisfactory in some circumstances (Peters, 1983).

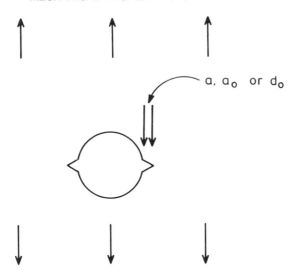

Figure 7.4. Illustrating the conceptual approach of Waddoups *et al.* (1971) and Whitney and Nuismer (1974) to account for the "hole size effect" in laminates. (Taken from Waddoups *et al.*, 1971.)

Potter (1978) has suggested that the hole size effect could be accounted for by the magnitude of the stress gradient in a laminate. It can be argued that the load carried by a fibre before fracture is transferred to the adjacent intact fibre. If this fibre is already carrying a high stress, the load transferred to it by the broken fibre can cause it to fail. Thus the possibility of sequential fibre failure depends on the initial stress gradient in the laminate. Using a theoretical analysis based on this argument it is possible to account for the observed decrease in strength with increasing hole size in $0°/\pm45°$ carbon fibre laminates. Yet another approach to the problem has been suggested by Caprino (1983), who proposed that the tensile strength σ_c of a notched laminate should be given by

$$\sigma_c = K_{1c}(\pi L)^{-m} \tag{7.8}$$

where $2L$ is the length of the notch in the laminate. It is then assumed that the strength σ_0 of a nominally undamaged laminate is due to intrinsic flaws of length L_0 so that

$$\frac{\sigma_c}{\sigma_0} = \left(\frac{L_0}{L}\right)^m \tag{7.9}$$

The values of L_0 and m can then be obtained from observations of the strengths of laminates containing notches of various lengths. Reasonable agreement with a wide range of data is obtained by taking m as approximately 0.3 and L_0 as approximately 0.2 mm.

The analysis of transverse crack growth described in §§4.4 and 4.6 can be modified to deal with the strength of damaged laminates (Morley, 1985a). In this case a crack is considered to extend through the matrix from the tip of the notch, becoming bridged by intact fibres, so that its propagation is hindered. The crack extends in a stable manner with increasing applied stress and this corresponds qualitatively with the observed progressive increase in size of the zone of damage. Eventually, catastrophic failure occurs when the stress carried by the crack-bridging fibres is sufficient to cause their progressive failure. The observed strength values of damaged laminates are predicted by the model, which also successfully predicts the strength of an unnotched lamina if adventitious damage of dimensions <0.1 mm is assumed to be present.

7.3 FATIGUE FAILURE

Fatigue failure can be defined as the eventual fracture of a structural material by the repeated application of a load that, if applied continuously, would be insufficient to cause failure. Generally, the repeated stress required to cause eventual failure by fatigue decreases monotonically as the number of loading cycles increases. The slope of the curve defining the magnitude of the repeated applied stress versus the number of cycles to failure thus provides a measure of the fatigue resistance of the material.

Both polymeric and metallic materials undergo fatigue failure by the initiation and incremental growth of cracks and are usually characterized by a limiting stress below which fatigue failure does not occur. Brittle elastic reinforcing fibres do not suffer strength degradation due to cyclic loading but they can be caused to fail at reduced stresses owing to chemical attack. The degradation of the strengths of glass fibres when stressed in the presence of water is an important example §2.1.1 that can occur also under cyclic loading.

The inhomogeneous nature of fibre composites leads to the generation of widely different stress levels within the material. For example, high local stresses are generated in the matrix and at the fibre–matrix interface when fibres are aligned at right-angles to the applied stress (§6.2), leading to interfacial debonding in these regions at low composite stress levels. Fibre composites generally contain voids and defects due to the limitations of the manufacturing process and these can act as sites for fatigue failure. A variety of failure modes may be present including fibre failure, interfacial debonding, delamination and matrix cracking. Fatigue failure in fibre composites proceeds in general by the accumulation of damage throughout the material. Fracture eventually occurs through the linking together of regions of damage at some cross-section of the material.

7.3.1   Fatigue of Unidirectional Systems

Reinforcing fibres have variable strengths owing to the presence of flaws. When a unidirectional composite is loaded in tension, in the direction of the fibre alignment, fibre failure occurs at random positions within the composite and the number of failures increase as the applied stress is increased.

When the matrix is a polymer, interfacial debonding will generally occur near the end of the fractured fibre and a transverse matrix crack may also propagate in this region. If the fibres are not degraded by cyclic loading and the cyclic strains are insufficient to generate fatigue effects in the matrix, then the composite will show little or no reduction in strength during cyclic loading. This condition is found, for example, in the case of a polymeric matrix reinforced with high-modulus type I carbon fibres.

Glass fibres have relatively low elastic moduli and high failing strains, so that matrix fatigue damage can occur when the peak cyclic strains developed in the composite are less than the fibre failure strains. The relatively high cyclic strains developed during fatigue loading of glass-fibre-reinforced polymers can cause the temperature of the material to increase owing to hysteresis losses. This effect is enhanced by the low thermal conductivity of the material and becomes noticeable at frequencies in excess of about 20 Hz. Such temperature increases result in a degradation of the mechanical properties of the matrix. Even at low loading frequencies, when temperature increases are negligible, the strengths of unidirectional glass-fibre-reinforced polymers show a pronounced fatigue effect, their strengths decreasing progressively when subjected to cyclic tensile loading.

Dharan (1976) has pointed out that the cyclic tensile strain experienced by the polymeric matrix of a fibre-reinforced composite should be less than the fatigue endurance limit of the matrix if long fatigue lives are to be achieved. This situation is illustrated in Figure 7.5, where the initial tensile strain developed in repeated tension cyclic loading is plotted against the number of loading cycles to failure for glass-fibre-reinforced epoxy resin, for high-modulus carbon-fibre-reinforced epoxy resin and for unreinforced epoxy resin.

Because of the low failing strain of high modulus carbon fibres the cyclic strains applied to this material are less than the fatigue endurance limit of the epoxy matrix at 10^6 cycles and no measurable degradation in strength of the composite is observed. In contrast, strength (plotted in terms of strain) of epoxy resin unidirectionally reinforced with glass fibres falls rapidly with increasing number of loading cycles and approaches that of the unreinforced matrix when the number of loading cycles is large. It should be emphasized that, because the elastic modulus of the carbon fibres is about five times that of the glass fibres, the short-term fatigue strengths of both types of composites are similar.

Talreja (1981) has pointed out that the fatigue characteristics of epoxy matrix composites utilizing type II and type III carbon fibres (plotted as cyclic strain values) fall between the curves for the glass fibre and carbon fibre composites shown in Figure 7.5. These fibres have failing strains intermediate between type I carbon fibres and glass fibres. Thus, when the fibres themselves are not subjected to fatigue failure, the fatigue life of a

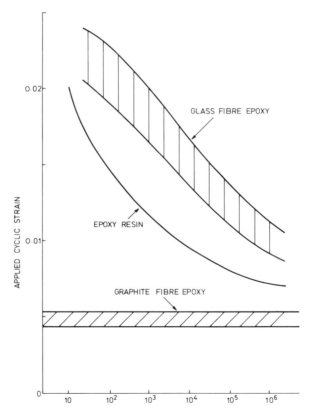

Figure 7.5. Applied cyclic strain versus cycles to failure. (Redrawn from Dharan (1976) with additional data from Talreja (1981).) Reprinted with permission from the *First International Conference on Composite Materials,* The American Institute of Mining Metallurgical and Petroleum Engineers, Inc., 345 East Seventh Street, New York, N.Y. 10017 U.S.A.

unidirectional polymer matrix composite is linked strongly to the fatigue characteristics of the matrix.

Similar behaviour has been noted by Morley and McColl (1981) and McColl and Morley (1982) using unidirectionally reinforced metal matrix composites with the reinforcing members pre-stressed in tension. Complete suppression of matrix fatigue crack growth was observed under zero-tension and tension-compression loading when the matrix was held in compression by the pre-stressed reinforcing members.

When the fibres in a unidirectionally reinforced composite are aligned transversely to a monotonically increasing tensile load it is observed that, for a variety of polymeric matrix composites, fibre matrix debonding is initiated at strains of about 0.3% (Owen, 1974). Under cyclic loading, the strain required to cause this type of damage is reduced as the number of loading cycles is increased, falling to about 0.1% at 10^6 cycles. Under cyclic loading, matrix cracks are initiated by fibre–matrix debonding, which thus sets a lower strain limit for the onset of fatigue damage.

If the fibres are aligned at an angle θ to the loading direction, mixed mode crack growth occurs during cyclic loading. Under repeated tensile loading, the fatigue strength of a unidirectionally reinforced composite decreases rapidly as θ is increased (Talreja, 1981).

When a unidirectional composite is loaded in axial compression, failure may take place when the fibres reach their compressive strengths or alternatively the fibres may buckle and fail in bending. The failure mode is governed by the mechanical properties of the fibres and their diameters, on the fibre matrix adhesion and matrix properties, and also on the presence of voids, notches and discontinuities in the composite structures. Carbon-fibre–polymer-matrix composites fail as a consequence of fibre buckling both under a monotonically increasing compressive load and also when subjected to cyclic compressive loading. For unidirectional high-modulus carbon-fibre–polymer-matrix composites both the compressive strength and the fatigue endurance stress under compressive loading are only about half of the corresponding tensile values. The fatigue strengths observed under fully reversed tensile and compressive stresses and under zero compression stress are indistinguishable, indicating the dominance of the compressive model of failure.

Fracture can be initiated from the surface of the material or from a void or notch. Cracks grow transversely through the composite structure by sequential fibre failure. The crack plane may change owing to local shear failure of the matrix (Prakash, 1981), and when fully reversed loading is applied failure eventually occurs in tension when the reduced cross-section of the material is no longer able to support the peak applied tensile load. When the material is loaded under zero-compression in flexure, failure is initiated at the compressive surface and a crack formed by successive fibre failure propagates towards the neutral axis. Under these loading conditions the fracture faces can continue to support a compressive load and the growth of the transverse crack is accompanied by longitudinal splitting of the material, which modifies the rate of growth of the primary transverse crack (Kunz and Beaumont, 1975).

Various studies of interlaminar shear fatigue have been made using specimens loaded in 3 and 4-point bending. These have been discussed by Owen (1974). Static shear stress values ranging from about $20\,\mathrm{MN\,m^{-2}}$ to about $70\,\mathrm{MN\,m^{-2}}$ can be obtained, depending on the type of fibres used, the fibre surface treatment and the form of the composite structure i.e. unidirectional or laminated. In general, the composites with the highest initial interlaminar shear strengths show the greatest proportional degradation of shear strength during cyclic loading.

7.3.2 Fatigue Characteristics of Laminates

The fatigue characteristics of laminates are very complex and a comprehensive treatment of them is beyond the scope of this book. A useful introduction to the extensive literature on this topic is provided by Lauraitis (1981) and Reifsnider (1982). In this section brief reference is made to some of the more important factors that influence the fatigue behaviour of these materials. The bulk of the investigations reported in the literature refer to unidirectional cyclic loading in tension-tension, tension-compression and compression-compression of carbon-fibre-reinforced epoxy resin laminates and most of the experimental observations quoted here refer to this type of material.

The fatigue failure processes occurring within the individual laminae depend on the basic characteristics of the lamina and the orientation of the fibres to the applied cyclic load. Failure of the interlaminar interfaces within a laminate can also occur and this modifies the constraint that one lamina exerts on the failure processes occurring in adjacent laminae. These effects are influenced strongly by the arrangement of the various laminae in a laminate. Cracks develop parallel to the direction of fibre alignment in the off-axis plies during cyclic loading. Their numbers increase up to some limiting value, which has been termed the characteristic damage state. The equilibrium crack spacing, eventually reached, depends on the ply angles and ply thicknesses, the stacking sequences of the various plies in the laminate and on the stress levels applied.

In addition to the growth of cracks parallel to the fibres within the off-axis plies during fatigue loading, delamination by the progressive growth of cracks between the individual plies also occurs. Delamination is initiated at the free edges of a laminate, since large inter-ply tensile and shear stresses are developed in these regions (§6.9). In addition, delamination of surface plies can be initiated by the prior development of through-ply cracks parallel to the fibres in these plies. The stress levels required to produce delamination again depend on the orientations, thicknesses and stacking sequences of the plies in the laminate, and delamination occurs between different plies at different stress levels.

As the area of delamination increases, and as the cracks form in the off-axis plies, the elastic modulus of the laminate falls and additional load is transferred to the 0° plies. These are more resistant to fatigue failure than the off-axis plies and control the long-term fatigue strength of $0°/\pm\theta°$ laminates. When the magnitude of the load transferred to them is sufficiently large, the 0° plies also fail in fatigue. The fatigue characteristics of a laminate thus depend on the relative proportions of the off-axis and 0° plies, their orientations and stacking sequences and on the fatigue characteristics of the 0° plies. Examples of these effects are shown by the behaviour of a 0°/90° cross-plied laminate containing high-elastic-modulus type I carbon fibres (Owen, 1974) and a $0°/\pm 45°$ laminate reinforced with relatively low-elastic-modulus carbon fibres (type III) (Sturgeon, 1977). The 90° plies in the 0°/90° laminate contribute little to its longitudinal strength, which is thus controlled by the fatigue-resistant low-extension 0° plies. The laminate thus shows little degradation of strength under fatigue loading. The contribution of the ±45° plies to the strength in the 0° direction of the $0°/\pm45°$ laminate is proportionally greater, so that the reduction in the strength of this laminate is proportionally greater as the ±45° plies become degraded during cyclic loading. Also, the higher elastic extension of the type III carbon fibres produces a greater reduction in the strength of the 0° plies during cyclic loading. This laminate thus shows a much greater reduction in strength under fatigue loading (Figure 7.6). Under compressive cyclic loading, the fatigue strength of a cross-ply laminate is governed by the plies that finally resist buckling and crushing. The compressive fatigue strength of a 0°/90° carbon fibre epoxy resin composite is governed by the compressive strength of the 0° plies and is much lower than the corresponding tensile fatigue properties. Data for this loading condition, given by Owen and Morris (1971), are shown in Figure 7.6. The fatigue characteristics of this particular laminate when subjected to zero compression loading are almost identical with those observed under fully reversed tension-compression loading. Also, the degradation of strength with increasing loading cycles is small, the fatigue strength in compression being little reduced from the static value. Generally the fatigue strengths of laminates subjected to cyclic tensile and compressive loading depend not only on the peak applied stress but on the ratio between the

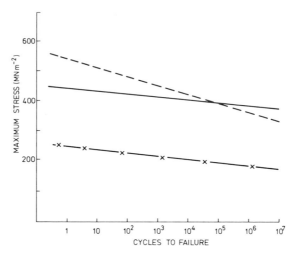

Figure 7.6. Typical strength values versus cycles to failure for 0°/±45° (type 3) carbon fibre laminates (– – – –). (Data from Sturgeon, 1977); also 0°/90° cross-ply laminates utilizing high modulus (type 1) carbon fibres (Owen and Morris, 1971; see also Owen, 1974.) Zero-tension loading indicated (———); zero-compression loading indicated (——— × ———). (Crown copyright RAE figure.)

maximum compressive and tensile stresses (Schutz and Gerharz, 1977). The fatigue characteristics also depend upon the ply stacking sequence.

From the above discussion it is apparent that the fatigue endurance strength, the residual strength and loss of stiffness after a given number of loading cycles will vary considerably according to the details of the laminate construction. They will also depend on the stress levels applied during cyclic loading, since different stress levels are required to develop the various types of degradation processes that can occur. In any practical application of fibre composites the effect of damage on the fatigue performance of the laminate has to be considered. This will usually result from damage by impact of foreign objects. Holes are produced at high impact velocities, while low impact velocities can produce local delamination between plies. The effect of fastener holes on the fatigue characteristics of laminates is also of importance.

Some general conclusions can be drawn from the observed fatigue characteristics of laminates containing notches or regions of damage. It is found that, in general, the tensile strengths of damaged laminates are not adversely affected by cyclic tensile loads but, conversely, their compressive strengths are significantly reduced by cyclic compressive loads. Moreover, compressive strengths under both static and cyclic loading can be impaired as a consequence of internal delamination caused by impact damage. When notched laminates are subjected to cyclic loading, cracks are developed between plies and within plies as a consequence of the enhanced stresses developed by the notch. These effects can enhance the residual tensile strength of a carbon-fibre–polymer-matrix laminate, since they relieve the localized high stresses developed by the notch in the 0° plies. However, this effect is not observed in all systems. For example, in quasi-isotropic boron-fibre–epoxy-resin laminates the residual strength after fatigue loading can be less than the initial notched strength (Roderick and Whitcomb, 1977), and no enhancement of the notched strength of quasi isotropic carbon fibre epoxy resin laminates was observed after repeated tensile loading by Walter *et al.* (1977). In the case of glass-fibre-reinforced 0°/90° laminates a transverse crack can propagate by the sequential failure in fatigue of ligaments of material formed ahead of the crack tip by the splitting of the laminate parallel to the loading direction and transversely to the crack axis (Mandell and Meier, 1975).

The static compressive strength of quasi-isotropic carbon-fibre–epoxy-resin laminates is reduced by the presence of a circular hole (Walter *et al.*, 1977). It is observed to fall further during fully reversed tension-compression cyclic loading and also under zero-compression cyclic loading, the strength under these conditions being governed by the peak compressive stress applied during cyclic loading. Quasi-isotropic carbon-fibre–epoxy-resin laminates are also sensitive to low-level impact damage when subsequently subjected to cyclic compressive loads. Impact energy levels that cause barely visible damage to the laminate are observed to reduce its strength considerably during subsequent compression fatigue loading. However, impact damage of this type does not necessarily impair the static tensile strength or the fatigue strength of the laminate in tension-tension loading. The effects of local delamination on the compressive fatigue strength of a laminate have been studied by the introduction of artificial delaminated zones between plies. The effect is observed to be dependent on the position of the delaminated zone and the lay-up of the laminate (Ramkumar 1982).

In the majority of engineering applications, structural materials are subjected to multi-axial fluctuating stresses. In §6.8 the generation of theoretical failure surfaces in stress space, as a means of predicting the conditions for failure under combined stresses, was discussed. It is reasonable to suppose that some guide to the combined stresses required to develop fatigue damage would be provided by similar surfaces in stress space. Various concentric surfaces of this type, corresponding to diminishing stress levels, would then represent the conditions for failure for increasing numbers of load cycles. The analytical problem is complicated by the uncertainty about the effects of interactions between different failure modes at different stress levels and loading cycles.

Systematic investigations requiring the production of large amounts of data are required to determine the dimensions of possible failure envelopes experimentally, but because of the complexity and cost of such studies, only limited experimental data are currently available. Two experimental techniques have been used. One makes use of flat cruciform test pieces, the combined loads being applied to the central region of the specimen through its four arms (see e.g. Pascoe and Tutton, 1982). The second technique makes use of tubular specimens, which are subjected to internal pressure and axial loads. A combination of axial loads and torsion can also be used (see §6.8). In many applications the reinforcement is in the form of a woven fabric and this leads to additional failure modes where the fibres cross over each other. The effectiveness of various theoretical failure envelopes to predict fatigue characteristics of composites under biaxial loading have been discussed by Owen and Rice (1982).

7.4 ENVIRONMENTAL EFFECTS

The mechanical properties of composites can be degraded in various ways by environmental effects. Some of the more important of these are discussed in outline in this section.

Because of the widespread use of carbon–fibre–epoxy-resin laminates in aircraft construction, considerable attention has been given in recent years to the effects of water absorption by these materials. Epoxy resins can absorb up to several per cent by weight of water. The equilibrium water content is controlled primarily by the relative humidity and the rate of moisture absorption increases with increasing temperature. These effects are almost completely reversible, the water being removed by heating in a vacuum or in a dry gaseous atmosphere. The absorbed water produces a volume increase in the resin and reduces its elastic moduli and glass transition temperature. Matrix cracks can be produced by non-uniform volume changes. In general, a volume change in the matrix produces an increase in the transverse dimensions of a unidirectional laminate, but the corresponding longitudinal expansion is inhibited by the reinforcing fibres. These effects can offset the stresses produced by differential thermal contraction between the fibres and the matrix after curing. The resin strength falls with increasing moisture content, although its failing strain may be increased (Browning et al., 1977). The presence of moisture would therefore be expected to affect primarily those properties of a laminate that are matrix-controlled. This is demonstrated by the behaviour of uni-directionally reinforced carbon-fibre–epoxy-resin composites in which the elastic moduli and strengths in transverse tensile and shear loading are reduced by the presence of moisture while the axial properties are unimpaired (Browning *et al.*, 1977). Similar effects to those caused by absorbed moisture are produced by increases in temperature. Again, the longitudinal tensile strength is insensitive to temperature effects up to about 100°C, while the matrix-controlled properties are reduced by such temperature increases. The effect of high environmental humidities and temperatures on the tensile properties of a laminate depend on the laminate construction, since only the off-axis plies are affected under these loading conditions.

The reduction in the strength of glass fibres under stress and in the presence of water has already been mentioned (§2.1.1). Much greater reductions in strength are observed when E glass reinforcing fibres are stressed in tension in the presence of mineral acids. These can diffuse through a polymeric matrix and cause stress corrosion of the fibres at low stress levels. Planar crack faces are produced by the sequential failure of the fibres at the tip of a propagating crack (Hogg and Hull, 1980). This effect is not encountered with glass fibres that are more chemically resistant (Hogg, 1983; Aveston and Sillwood, 1982) and is limited in practice by the large factors of safety normally used in the design of chemical process plant (Roberts, 1982).

References

Adams, D. F. and Doner, D. R. (1967a). *J. Composite Mater.* **1,** 4–17.
Adams, D. F. and Doner, D. R. (1967b). *J. Composite Mater.* **1,** 152–164.
Adler, R. P. I. and Hammond, M. L. (1969). *Appl. Phys. Lett.* **14,** 354–358.
Allen, S., Cooper, G. A., Johnson, D. J. and Mayer, R. M. (1970). *Third Conference on Industrial Carbons and Graphite*, pp. 456–462. Society of Chemical Industry, London.
Andersson, C.-H. and Warren, R. (1984). *Composites* **15,** 16–24.
Andrews, R. D. and Kimmel, R. M. (1965). *Polymer Lett.* **3,** 167–169.
Argon, A. S. (1972). In *Treatise on Materials Science and Technology* (H. Herman, ed.), Vol. 1, pp. 79–114. Academic Press, London and Orlando.
Arridge, R. G. C. and Heywood, D. (1967). *Br. J. Appl. Phys.* **18,** 447–457.
Ashton, J. E., Halpin, J. C. and Petit, P. H. (1969). *Primer on Composite Materials.* Technomic Publishing Company, Stamford, Connecticut.
Aveston, J. and Kelly, A. (1973). *J. Mater. Sci.* **8,** 352–362.
Aveston, J. and Kelly, A. (1980). *Phil. Trans. Roy. Soc. Lond.* **A294,** 519–534.
Aveston, J. and Sillwood, J. M. (1976). *J. Mater. Sci.* **11,** 1877–1883.
Aveston, J. and Sillwood, J. M. (1982). *J. Mater. Sci.* **17,** 3491–3498.
Aveston, J., Cooper, G. A. and Kelly, A. (1971). In *Properties of Fibre Composites*, Conference Proceedings National Physical Laboratory, pp. 15–25. IPC Science and Technology Press, Guildford.
Bacon, R. (1973). In *Chemistry and Physics of Carbon* (P. L. Walker Jr. and P. A. Thrower, eds), Vol. 9, pp. 1–102. Marcel Dekker Inc., New York.
Bacon, R. (1979). *Phil. Trans. Roy. Soc. Lond.* **A294,** 437–442.
Bacon, R. and Tang, M. M. (1964). *Carbon* **2,** 221–225.
Bailey, J. E., Curtis, P. T. and Parvizi, A. (1979). *Proc. Roy. Soc. Lond.* **A366,** 599–623.
Baker, C., Bentall, R. H., Hawthorne, H. and Linger, K. R. (1970). In *Third*

Conference on Industrial Carbons and Graphite, pp. 508–510. Society of Chemical Industry, London.

Barry, P. W. (1978). *J. Mater. Sci.* **13**, 2177–2187.

Bartos, P. (1980). *J. Mater. Sci.* **15**, 3122–3128.

Batdorf, S. B. (1982). *J. Reinforced Plastics and Composites* **1**, 153–164.

Batdorf, S. B. and Ghaffarian, R. (1982). *J. Mater. Sci. Lett.* **1**, 295–297.

Beaumont, P. W. R. and Harris, B. (1971). In *Carbon Fibres: Their Composites and Applications*, pp. 283–291. The Plastics Institute, London.

Bennett, S. C. and Johnson, D. J. (1978). *Fifth London International Carbon and Graphite Conference*, pp. 377–386. Society of Chemical Industry, London.

Bennett, S. C., Johnson, D. J. and Johnson, W. (1983). *J. Mater. Sci.* **18**, 3337–3347.

Blackman, L. C. K., Proctor, B. A., Smith, J. W. and Taylor, J. W. (1977). *The Chartered Mechanical Engineer* **24** (No. 1), 45–51.

Bowling, J. and Groves, G. W. (1979). *J. Mater. Sci.* **14**, 431–442.

Bromley, J. (1971). In *Carbon Fibres: Their Composites and Applications*, pp. 3–9. The Plastics Institute, London.

Brooks, J. D. and Taylor, G. H. (1965). *Carbon* **3**, 185–193.

Brooks, J. D. and Taylor, G. H. (1968). In *Chemistry and Physics of Carbon* (P. L. Walker, Jr., ed.), vol. 4, pp. 243–286. Edward Arnold, London; Marcel Dekker Inc., New York.

Browning, C. E., Husman, G. E. and Whitney, J. M. (1977). In *Composite Materials Testing and Design, Fourth Conference*, pp. 481–496. Special Technical Publication 617. American Society for Testing and Materials, Philadelphia.

Bunsell, A. R. and Harris, B. (1974). *Composites* **5**, 157–164.

Caprino, G. (1983). *J. Mater. Sci.* **18**, 2269–2273.

Carlsson, J.-O. (1979). *J. Mater. Sci.* **14**, 255–264.

Chaplin, C. R. (1977). *J. Mater. Sci.* **12**, 347–352.

Chappell, M. J., Morley, J. G. and Martin, A. (1975). *J. Phys. D. Appl. Phys.* **8**, 1071–1083.

Charles, R. J. (1958). *J. Appl. Phys.* **29**, 1549–1560.

Chen, S. S., Herms, J., Peebles, L. H. and Uhlmann, D. R. (1981). *J. Mater. Sci.* **16**, 1490–1510.

Cheng, D. C. and Weng, G. W. (1979). *J. Mater. Sci.* **14**, 2183–2190.

Chua, P. S. and Piggott, M. R. (1985). *Composites Sci. Technol.* **22**, 33–42.

Clark, D., Wadsworth, N. J. and Watt, W. (1974). In *Carbon Fibres; Their Place in Modern Technology*, pp. 44–51. The Plastics Institute, London.

Clarke, A. J. and Bailey, J. E. (1974). In *Carbon Fibres, Their Place in Modern Technology*, pp. 12–15. The Plastics Institute, London.

Coleman, B. D. (1958). *J. Mech. Phys. Solids* **7**, 60–70.

Collings, T. A. (1974). *Composites* **5**, 108–116.

Cook, J. (1968). *Br. J. Appl. Phys. Ser. 2.* **1**, 799–812.

Cook, J. and Gordon, J. E. (1964). *Proc. Roy. Soc. Lond.* **A282**, 508–520.

Cooper, G. A. (1971). *Rev. Phys. Technol.* **2**(No. 2), 49–91.

Cooper, G. A. and Kelly, A. (1967). *J. Mech. Phys. Solids* **15**, 279–297.

Cooper, G. A. and Sillwood, J. M. (1972). *J. Mater. Sci.* **7**, 325–333.

Cottrell, A. H. (1964). *The Mechanical Properties of Matter*. John Wiley and Sons Inc., New York, London and Sydney.

Cox, H. L. (1952). *Br. J. Appl. Phys.* **3**, 72–79.

Curtis, P. T. (1984). *J. Mater. Sci.* **19**, 167–182.

Daniel, I. M. (1974). In *Mechanics of Composite Materials* (G. P. Sendeckyj, ed.), pp. 433–489. Vol. 2 of *Composite Materials* (L. J. Broutman and R. H. Krock, eds). Academic Press, London and Orlando.

DaSilva, J. L. G. and Johnson, D. J. (1984). *J. Mater. Sci.* **19**, 3201–3210.

Deteresa, S. J., Allen, S. R., Farris, R. J. and Porter, R. S. (1984). *J. Mater. Sci.* **19**, 57–72.

Dharan, C. K. H. (1976) In *First International Conference on Composite Materials*, Vol. 1, pp. 830–839. The American Institute of Mining, Metallurgical and Petroleum Engineers Inc., New York.

Dhingra, A. K. (1980). *Phil. Trans. Roy. Soc. Lond.* **A294**, 411–417.

Dingle, L. E. (1974). In *Carbon Fibres: Their Place in Modern Technology*, pp. 78–86. The Plastics Institute, London.

Divecha, A. P., Fishman, S. G. and Karmarkar, S. D. (1981). *J. Metals* **33**, 12–17.

Dobb, M. G., Johnson, D. J. and Saville, B. P. (1979). *Phil. Trans. Roy. Soc. Lond.* **A294**, 483–485.

Dow, N. F. (1963). General Electric Company Report, R 63SD61.

Edwards, H. and Evans, N. P. (1980). In *Advances in Composite Materials, Proceedings of the Third International Conference on Composite Materials, Paris*, (A. R. Bunsell, C. Bathias, A. Martrenchar, D. Menkes and G. Verchery, eds), Vol. 2, pp. 1620–1635. Pergamon Press, Oxford.

Edwards, H. E., Parratt, N. J. and Potter, K. D. (1978). In *Proceedings of the Second International conference on Composite Materials* (B. Noton, R. Signorelli, K. Street and L. Phillips, eds), pp. 975–990. The American Institute of Mining, Metallurgical and Petroleum Engineers Inc., New York.

Ehrburger, P. and Donnet, J. B. (1980). *Phil. Trans. Roy. Soc. Lond.* **A294**, 495–505.

Flaggs, D. L. and Kural, M. H. (1982). *J. Composite Mater.* **16**, 103–116.

Fuwa, M., Bunsell, A. R. and Harris, B. (1975). *J. Mater. Sci.* **10**, 2062–2070.

Galasso, F., Salkind, M., Kuehl, D. and Patarini, V. (1966). *Trans AIME* **236**, 1748–1751.

Garrett, K. W. and Bailey, J. E. (1977). *J. Mater. Sci.* **12**, 157–168.

Goodhew, P. J., Clarke, A. J. and Bailey, J. E. (1975). *Mater. Sci. Eng.* **17**, 3–30.

Grassie, N. and Hay, J. N. (1962). *J. Polymer Sci.* **56**, 189–202.

Gray, R. J. (1984). *J. Mater. Sci.* **19**, 861–870.

Greenwood, J. H. (1977). *Composites* **8**, 175–184.

Griffith, A. A. (1920). *Phil Trans. Roy. Soc. Lond.* **A221**, 163–198.

Halpin, J. C. and Tsai, S. W. (1969). In an appendix of *Environmental Factors in Composite Materials*. Air Force Materials Laboratory Report TR 67-423.

Harlow, D. G. and Phoenix, S. L. (1978). *J. Composite Mater.* **12**, 195–214.

Harlow, D. G. and Phoenix, S. L. (1981). *Int. J. Fracture* **17**, 601–630.

Harris, B. (1978). *J. Mater. Sci.* **13**, 173–177.

Harris, B. and Ankara, A. O. (1978). *Proc. Roy. Soc. Lond.* **A359**, 229–250.

Harris, B., Morley, J. and Phillips, D. C. (1975). *J. Mater. Sci.* **10**, 2050–2061.

Hasegawa, Y. and Okamura, K. (1983). *J. Mater. Sci.* **18**, 3633–3648.

Hasegawa, Y., Iimura, M. and Yajima, S. (1980). *J. Mater. Sci.* **15**, 720–728.

Hashin, Z. (1970). In *Mechanics of Composite Materials* (F. W. Wendt, H. Leibowitz and N. Perrone, eds), pp. 201–241. Pergamon Press, Oxford.

Hawthorne, H. M. (1976). *J. Mater. Sci.* **11**, 97–110.

Hay, J. N. (1968). *J. Polymer Sci.* **A-1, 6**, 2127–2135.

Higuchi, S., Otsuka, R. and Shiraishi, M. (1984). *J. Mater. Sci.* **19**, 270–278.

Hill, R. (1950). *The Mathematical Theory of Plasticity*. Oxford University Press, London.
Hill, R. (1964). *J. Mech. Phys. Solids* **12**, 199–212; 213–218.
Hillig, W. B. (1962). In *C. R. symposium sur la resistance mechanique du verre et les moyens de l'ameliorer*, pp. 295–325. Union Scientifique Continentale du Verre.
Hogg, P. J. (1983). *Composites* **14**, 254–261.
Hogg, P. J. and Hull, D. (1980). *Metal Science* **17**, 441–449.
Hollister, G. S. and Thomas, C. (1966). *Fibre Reinforced Materials*. Elsevier, London.
Hull, D., Legg, M. J. and Spencer, B. (1978). *Composites* **9**, 17–24.
Inglis, C. E. (1913). *Trans. Inst. Naval Archit.* **55**, 219–230.
Irwin, G. R. (1957). *J. Appl. Mech.* **24**, 361–364.
Ishikawa, T. and Chou, T. E. (1982). *J. Mater. Sci.* **17**, 3211–3220.
Jackson, P. W. and Cratchley, D. (1966). *J. Mech. Phys. Solids* **14**, 49–64.
Jacob, B. A., Douglas, F. C. and Galasso, F. S. (1973). *Amer. Ceram. Soc. Bull.* **52**, 896–897.
Jain, M. K. and Abhiraman, A. S. (1983). *J. Mater. Sci.* **18**, 179–188.
Johnson, J. W. (1983). Composites **14**, 107–114.
Johnson, J. W., Marjoram, J. R. and Rose, P. G. (1969). *Nature* **221**, 357–358.
Johnson, J. W., Rose, P. G. and Scott, G. (1970). In *Third Conference, Industrial Carbons and Graphite*, pp. 443–446. Society of Chemical Industry, London.
Johnson, W. (1970). In *Third Conference, Industrial Carbon and Graphite*, pp. 447–452. Society of Chemical Industry, London.
Johnson, W. (1979). *Nature* **279**, 142–145.
Johnson, W., Phillips, L. N. and Watt, W. (1964). U.K. Patent No. 1,110,791.
Jones, R. M. (1975). *Mechanics of Composite Materials*. Scripta Book Company, Washington D.C. (McGraw Hill Book Company, New York).
Jones, M. C. L. and Hull, D. (1979). *J. Mater. Sci.* **14**, 165–174.
Jones, J. B., Barr, J. B. and Smith, R. E. (1980). *J. Mater. Sci.* **15**, 2455–2465.
Jones, W. R. and Johnson, J. W. (1971). *Carbon* **9**, 645–655.
Kelly, A. (1970). *Proc. Roy. Soc. Lond.* **A319**, 95–116.
Kelly, A. (1973). *Strong Solids*. (2nd edition) Oxford University Press, Oxford.
Kelly, A. and Davis, G. J. (1965). *Metall. Rev.* **10**, 1–77.
Kelly, A. and Zweben, C. (1976). *J. Mater. Sci. Lett.* **11**, 582–587.
Kelsey, R. H. (1967). In *Modern Composite Materials* (L. J. Broutman and R. H. Krock, eds), pp. 217–227. Addison Wesley, Reading, Mass.
Kies, J. A. (1962). *Maximum strains in the resin of fiberglass composites*. U.S. Naval Research Laboratory Report NRL 5752.
Kirk, J. N., Munro, M. and Beaumont, P. W. R. (1978). *J. Mater. Sci.* **13**, 2197–2204.
Korczynskyj, Y. and Morley, J. G. (1981). *J. Mater. Sci.* **16**, 1785–1795.
Korczynskyj, Y., Harris, S. J. and Morley, J. G. (1981). *J. Mater. Sci.* **16**, 1533–1547.
Kreider, K. G. and Prewo, K. M. (1972). Special Technical Publication No. 497, pp. 539–550. American Society for Testing and Materials, Philadelphia.
Kreider, K. G. and Prewo, K. M. (1974). In *Composite Materials* (L. J. Broutman and R. H. Krock, eds), Vol. 4, Metallic Matrix Composites, pp. 399–471. Academic Press, New York and Orlando.
Kunz, S. C. and Beaumont, P. W. R. (1975). In *Fatigue of Composite Materials*, pp. 71–91. Special Technical Publication 569. American Society for Testing and Materials, Philadelphia.

Lauratis, K. N. (Symposium chairman) (1981). *Fatigue of Fibrous Composite Materials*. Special Technical Publication 723. American Society for Testing and Materials, Philadelphia.

Lee, J.-G. and Cutler, I. B. (1975). *Ceram. Bull.* **54,** 195–198.

Legg, M. J. and Hull, D. (1982). *Composites* **13,** 369–376.

Litherland, K. L., Oakley, D. R. and Proctor, B. A. (1981). *Cement and Concrete Res.* **11,** 455–466.

Loewenstein, K. L. (1973). *The Manufacturing Technology of Continuous Glass Fibres.* Elsevier Scientific Publishing Company, Amsterdam, London and New York.

Love, G., Cox, M. G. C. and Scott, V. D. (1975). *Mater. Res. Bull.* **10,** 815–818.

Lowrie, R. E. (1967). In *Modern Composite Materials* (L. J. Broutman and R. H. Krock, eds), pp. 270–323. Addison Wesley, Reading, Mass.

McColl, I. R. and Morley, J. G. (1977a). *Phil Trans. Roy. Soc. Lond.* **A287,** 17–43.

McColl, I. R. and Morley, J. G. (1977b). *J. Mater. Sci.* **12,** 1165–1175.

McColl, I. R. and Morley, J. G. (1982). *Int. J. Fracture.* **18,** 191–216.

McGarry, F. J. and Mandell, J. F. (1972). In *27th Annual Technical Conference Reinforced Plastics/Composites Institute Section 9A*, pp. 1–12. The Society of the Plastics Industry, Inc.

Magat, E. E. (1980). *Phil. Trans. Roy. Soc. Lond.* **A294,** 463–472.

Mah, T., Hecht, N. L., McCullum, D. E., Hoenigman, J. R., Kim, H. M., Katz, A. P. and Lipsitt, H. A. (1984). *J. Mater. Sci.* **19,** 1191–1201.

Majumdar, A. J. (1970). *Proc. Roy. Soc. Lond.* **A319,** 69–78.

Mallinder, F. P. and Proctor, B. A. (1964). *Phys. Chem. Glasses* **5,** 91–103.

Mandell, J. F. and Meier, U. (1975). In *Fatigue of Composite Materials*, pp. 28–44. Special Technical Publication 569. American Society for Testing and Materials, Philadelphia.

Mandell, J. F., Wang, Su-Su and McGarry, F. J. (1975). *J. Composite Mater.* **9,** 266–287.

Manders, P. W., Chou, Tsu-Wei, Jones, F. R. and Rock, J. W. (1983). *J. Mater. Sci.* **18,** 2876–2889.

Markham, M. F. and Dawson, D. (1975). *Composites* **6,** 173–176.

Marsh, D. M. (1964). *Proc. Roy. Soc. Lond.* **A282,** 33–43.

Martinez, G. M., Piggot, M. R., Bainbridge, D. M. R. and Harris, B. (1981). *J. Mater. Sci.* **16,** 2831–2836.

Mehan, R. L. (1978). *J. Mater. Sci.* **13,** 358–366.

Mehan, R. L. and Noone, N. J. (1974). In *Composite Materials* (L. J. Broutman and R. H. Krock, eds.). Vol. 4, pp. 159–227. Academic Press, London and Orlando.

Mileiko, S. T. (1969). *J. Mater. Sci.* **4,** 974–977.

Milewski, J. V., Gac, F. D., Petrovic, J. J. and Skaggs, S. R. (1985). *J. Mater. Sci.* **20,** 1160–1166.

Millman, R. S. and Morley, J. G. (1975). *J. Phys. D: Appl. Phys.* **8,** 1065–1070.

Morley, J. G. (1964). *Proc. Roy. Soc. Lond.* **A282,** 43–52.

Morley, J. G. (1965). *Glass Technol.* **6,** 69–89.

Morley, J. G. (1971). *Carbon Fibers.* Kirk-Othmer Encyclopedia of Chemical Technology, Supplement volume 2nd edition, pp. 109–120. John Wiley and Sons, Inc., New York.

Morley, J. G. (1983). *J. Mater. Sci.* **18,** 1564–1576.

Morley, J. G. (1985a). *J. Mater. Sci.* **20,** 1794–1806.

Morley, J. G. (1985b). *J. Mater. Sci.* **20,** 3939–3948.

Morley, J. G. and McColl, I. R. (1981). *Engineering Fracture Mechanics*, **14,** 1–25.

Morley, J. G. and McColl, I. R. (1984). *J. Mater. Sci.* **19**, 3407–3415.
Morley, J. G. and Millman, R. S. (1974). *J. Mater. Sci.* **9**, 1171–1182.
Morley, J. G. Andrews, P. W. and Whitney, I. (1964). *Phys. Chem. Glasses* **5**, 1–10.
Morley, J. G., Millman, R. S. and Martin, A. (1976). *J. Phys. D: Appl. Phys.* **9**, 1031–1047.
Moreton, R. (1971). In *Carbon Fibres: Their Composites and Applications*, pp. 73–80. The Plastics Institute, London. (See also Moreton R. and Watt, W. (1974). *Carbon* **12**, 543–554.)
Morton, J. and Groves, G. W. (1974). *J. Mater. Sci.* **9**, 1436–1445.
Müller, D. J., Fitzer, E. and Fielder, A. K. (1971). In *Carbon Fibres, Their Composites and Applications*, pp. 10–17. The Plastics Institute, London.
Nicholas, J. and Ashbee, K. H. G. (1978). *J. Phys. D. Appl. Phys.* **11**, 1015–1017.
Ochiai, S. and Peters, P. W. M. (1982a). *J. Mater. Sci.* **17**, 417–418.
Ochiai, S. and Peters, P. W. M. (1982b). *J. Mater. Sci.* **17**, 2324–2336.
Okamura, K., Sato, M. and Hasegawa, Y. (1983). *J. Mater. Sci. Lett.* **2**, 769–771.
Orowan, E. (1949). *Rep. Prog. Phys.* **12**, 185–232.
Otte, H. M. and Lipsitt, H. A. (1966). *Phys. Stat. Solidi* **13**, 439–448.
Owen, M. J. (1974). In *Composite Materials* (L. J. Broutman and R. H. Krock, eds), Vol. 5, Fracture and Fatigue (L. J. Broutman ed.), pp. 313–369. Academic Press, London and Orlando.
Owen, M. J. and Morris, S. (1971). In *Carbon Fibres: Their Composites and Applications*. Plastics and Polymers Conference Supplement No. 5. Proceedings of the International Conference. The Plastics Institute, London.
Owen, M. J. and Rice, D. J. (1981). *Composites* **12**, 13–25.
Owen, M. J. and Rice, D. J. (1982). In *Composite Materials Testing and Design, Sixth Conference* (I. M. Daniel, ed.), pp. 124–143. Special Technical Publication 787. American Society for Testing and Materials, Philadelphia.
Pagano, N. J. and Pipes, R. B. (1971). *J. Composite Mater.* **5**, 50–57.
Paris, P. C. and Sih, G. C. (1965). In *Fracture Toughness Testing and its Applications*, pp. 30–81. Special Technical Publication 381. American Society for Testing and Materials, Philadelphia.
Parratt, N. J. (1966). *Chemical Engineering Progress* **62**, (No. 3), 61–67.
Parratt, N. J. and Potter, K. D. (1980). In *Advances in Composite Materials*, Proceedings of the Third International Conference on Composite Materials (A. R. Bunsell, C. Bathias, A. Martrenchar, D. Menkes and G. Verchery, eds), Vol. 1, pp. 313–326. Pergamon Press, Oxford.
Parry, T. V. and Wronsky, A. S. (1981). *J. Mater. Sci.* **16**, 439–450.
Parry, T. V. and Wronsky, A. S. (1982). *J. Mater. Sci.* **17**, 893–900.
Parvizi, A. and Bailey, J. E. (1978). *J. Mater. Sci.* **13**, 2131–2136.
Pascoe, K. J. and Tutton, P. A. (1982). In *Fatigue and Creep of Composite Materials* (H. Lilholt and R. Talreja, eds), pp. 265–270. Proceedings of the 3rd Riso International Symposium on Metallurgy and Materials Science. Riso National Laboratory, Roskilde, Denmark.
Peters, P. W. M. (1983). *Composites* **14**, 365–369.
Petrovic, J. J., Milewski, J. V., Rohr, D. L. and Gac, F. D. (1985). *J. Mater. Sci.* **20**, 1167–1177.
Phillips, D. C. and Wells, G. M. (1982). *J. Mater. Sci. Lett.* **1**, 321–324.
Piggot, M. R. (1981). *J. Mater. Sci.* **16**, 2837–2845.
Piggot, M. R. and Harris, B. (1980). *J. Mater. Sci.* **15**, 2523–2538.

Piggott, M. R. and Harris, B. (1981). *J. Mater. Sci.* **16**, 687–693.
Piggott, M. R. and Wilde, P. (1980). *J. Mater. Sci.* **15**, 2811–2815.
Pipes, R. B. and Cole, B. W. (1973). *J. Composite Mater.* **7**, 246–256.
Pipes, R. B. and Pagano, N. J. (1970). *J. Composite Mater.* **4**, 538–548.
Plueddmann, E. P. (ed.) (1974). *Composite Materials*, Vol. 6, Interfaces in Polymer Matrix Composites (general eds. L. J. Broutman and R. H. Krock). Academic Press, London and Orlando.
Potter, R. T. (1978). *Proc. Roy. Soc. Lond.* **A361**, 325–341.
Prakash, R. (1981). *Fibre Sci. Technol.* **14**, 171–181.
Prewo, K. M. (1974). *J. Composite Mater.* **8**, 411–414.
Prewo, K. M. and Kreider, K. G. (1972). *J. Composite Mater.* **6**, 338–357.
Proctor, B. A. (1962). *Phys. Chem. Glasses* **3**, 7–27.
Proctor, B. A. and Yale, B. (1980). *Phil. Trans. Roy. Soc. Lond.* **A294**, 427–436.
Proctor, B. A., Whitney, I. and Johnson, J. W. (1967). *Proc. Roy. Soc. Lond.* **A297**, 534–557.
Puck, A. and Schneider, W. (1969). *Plastics and Polymers* **37**, 33–44.
Purslow, D. (1981). *Composites* **12**, 241–247.
Ramkumar, R. L. (1982). In *Damage in Composite Materials* (K. L. Reifsnider, ed.), pp. 184–210. Special Technical Publication 775. American Society for Testing and Materials, Philadelphia.
Reifsnider, K. L. (ed.) (1982). *Damage in Composite Materials*. Special Technical Publication 775. American Society for Testing and Materials, Philadelphia.
Reynolds, W. N. and Moreton, R. (1980). *Phil. Trans. Roy. Soc. Lond.* **A294**, 451–461.
Reynolds, W. N. and Sharp, J. V. (1974). *Carbon* **12**, 103–110.
Roberts, J. R. (1967). In *Modern Composite Materials* (L. J. Broutman and R. H. Krock, eds), pp. 228–243. Addison Wesley, Reading, Mass.
Roberts, R. C. (1982). *Composites* **13**, 389–392.
Roderick, G. L. and Whitcomb, J. D. (1977). In *Fatigue of Filamentary Composite Materials* (K. L. Reifsnider and L. N. Lauraitis, eds), pp. 73–88. Special Technical Publication 636. American Society for Testing and Materials, Philadelphia.
Rose, P. G. (1971). Ph.D. Thesis, University of Aston in Birmingham, U.K.
Rosen, B. W. (1964). *AIAA Journal* **2**, 1985–1991.
Rosen, B. W. (1965). In *Fiber Composite Materials* (S. H. Bush, ed.), pp. 37–75. American Society for Metals, Metals Park, Ohio.
Ross, S. E. (1968). *Textile Res. J.* **38**, 906–913.
Ruland, W. (1969). *Appl. Polymer Symp.* **9**, 293–301.
Sanders, R. E. (1968). *Chemical and Process Engineering* **49**, 100–109.
Schutz, D. and Gerharz, J. J. (1977). *Composites* **8**, 245–250.
Sharp, J. V., Burnay, S. G., Matthews, J. R. and Harper, E. A. (1974). In *Carbon Fibres: Their Place in Modern Technology*, pp. 25–31. The Plastics Institute, London.
Shindo, A. (1961). Report No. 317, *Studies on Graphite Fiber*. Government Industrial Research Institute, Osaka, Japan.
Shindo, A. (1971). In *Carbon Fibres Their Composites and Applications*, pp. 18–22. The Plastics Institute, London.
Signorelli, R. A. (1974). In *Composite Materials* (L. J. Broutman and R. H. Krock, eds), Vol. 4, pp. 230–267. Academic Press, London and Orlando.
Simon, G. and Bunsell, A. R. (1984a). *J. Mater. Sci.* **19**, 3649–3657.

Simon, G. and Bunsell, A. R. (1984b). *J. Mater. Sci.* **19**, 3658–3670.

Sinclair, J. H. and Chamis, C. C. (1979). *Fracture Modes in Off-Axis Fiber Composites*, Proceedings of the 34th SPI/RP Annual Conference, Paper 22-A. Society of Plastics Industry, New York.

Smith, R. L., Phoenix, S. L., Greenfield, M. R., Henstenburg, R. B. and Pitt, R. E. (1983). *Proc. Roy. Soc. Lond.* **A388**, 353–391.

Sokolnikoff, I. S. (1956). *Mathematical Theory of Elasticity*, second edition. McGraw Hill Book Company, New York.

Standage, A. E. and Prescott, R. (1965). U.K. Patent No. 1,170,591.

Standage, A. E. and Prescott, R. (1966). *Nature* **211**, 169–169. (See also U.K. Patent No. 1,128,043.)

Sturgeon, J.B. (1977). *Composites* **8**, 221–226.

Takaku, A. and Arridge, R. G. C. (1973). *J. Phys. D: Appl. Phys.* **6**, 2038–2047.

Talley, C. P. (1959). *J. Appl. Phys.* **30**, 1114–1115. (See also U.S. Patent 3,491,055.)

Talreja, R. (1981). *Proc. Roy. Soc. Lond.* **A378**, 461–475.

Tang, M. M. and Bacon, R. (1964). *Carbon* **2**, 211–220.

Tattersall, H. G. and Tappin, G. (1966). *J. Mater. Sci.* **1**, 296–301.

Thomas, W. F. (1960). *Phys. Chem. Glasses* **1**, 4–18.

Thompson, E. F. and Lemkey, F. D. (1974). In *Composite Materials* (L. J. Broutman and R. H. Krock, eds), Vol. 4, Metallic Matrix Composites, pp. 101–157. Academic Press, London and Orlando.

Thorne, D. J. (1974). *Nature* **248**, 754–756.

Timoshenko, S. and Goodier, (1951). *Theory of Elasticity*, second edition. McGraw Hill Book Company, New York.

Tirosh, J., Katz, E., Lifschuetz, G. and Tetelman, A. S. (1979). *Engineering Fracture Mechanics* **12**, 267–277.

Tsai, S. W. (1968). In *Fundamental Aspects of Fiber Reinforced Plastic Composites* (R. T. Schwartz and H. S. Schwartz, eds), pp. 3–11. Wiley Interscience, New York.

Tsai, S. W. and Hahn, H. T. (1980). *Introduction to Composite Materials*. Technomic Publishing Company Inc., Westport, Connecticut.

Tsai, S. W. and Wu, E. W. (1971). *J. Composite Mater.* **5**, 58–80.

Tyson, C. N. (1975). *J. Phys. D: Appl. Phys.* **8**, 749–758.

Vega-Boggio, J. and Vingsbo, O. (1977). *J. Mater. Sci.* **12**, 2519–2524.

Vega-Boggio, J., Vingsbo, O. and Carlsson, J. O. (1977). *J. Mater. Sci.* **12**, 1750–1758.

Veltri, R. and Galasso, F. (1968). *Nature* **220**, 781–782.

Vinson, J. R. and Chou, T. W. (1975). *Composite Materials and Their Structures*. Applied Science Publishers, Barking.

Volk, H. F. (1977). *High Modulus Pitch-Based Carbon Fibres*. Union Carbide Corporation, Parma, Ohio.

Vosburgh, W. G. (1960). *Textile Res. J.* **30**, 882–896.

Waddoups, M. E., Eisemann, J. R. and Kaminski, B. E. (1971). *J. Composite Mater.* **5**, 446–454.

Wadsworth, N. J. and Spilling, I. (1968). *Br. J. Appl. Phys. (J. Phys. D.) Ser. 2*, **1**, 1049–1058.

Walter, R. W., Johnson, R. W., June, R. R. and McCarty, J. E. (1977). In *Fatigue of Filamentary Composite Materials* (K. L. Reifsnider and L. N. Lauraitis, eds), pp. 228–247. Special Technical Publication 636. American Society for Testing and Materials, Philadelphia.

Warren, R. and Andersson, C-H. (1984). *Composites* **15** (No. 2), 101–111.

Warner, S. B., Peebles, L. H. and Uhlmann, D. R. (1979a). *J. Mater Sci.* **14,** 556–564.

Warner, S. B., Peebles, L. H. and Uhlmann, D. R. (1979b). *J. Mater. Sci.* **14,** 1893–1900.

Warner, S. B., Peebles, L. H. and Uhlmann, D. R. (1979c). *J. Mater. Sci. Lett.* **14,** 2764–2765.

Watt, W. (1970). *Proc. Roy. Soc. Lond.* **A319,** 5–15.

Watt, W. and Johnson, W. (1970). In *Third Conference, Industrial Carbon and Graphite*, pp. 417–426. Society of Chemical Industry, London.

Watt, W., Johnson, D. J., and Parker, E. (1974). In *Carbon Fibres: Their Place in Modern Technology*, pp. 3–10. The Plastics Institute, London.

Wawner, F. E. (1967). In *Modern Composite Materials* (L. J. Broutman and R. H. Krock, eds), pp. 244–269. Addison Wesley, Reading, Mass.

Wawner, F. E. and Satterfield, D. B. (1967). *Appl. Phys. Lett.* **11,** 192–194.

Weaver, C. W. and Williams, J. G. (1975). *J. Mater. Sci.* **10,** 1323–1333.

Welles, J. K. and Beaumont, P. W. R. (1982). *J. Mater. Sci.* **17,** 397–405.

Whitney, I. (1973). *Composites* **4,** 101–104.

Whitney, J. M. and Nuismer, R. J. (1974). *J. Composite Mater.* **8,** 253–265.

Wronski, A. S. and Parry, T. V. (1982). *J. Mater. Sci.* **17,** 3656–3622.

Yajima, S., Hasegawa, Y., Hayashi, J. and Iimura, M. (1978). *J. Mater. Sci.* **13,** 2569–2576.

Zachariasen, W. H. (1932). *J. Amer. Chem. Soc.* **54,** Part 3, 3841–3851.

Zweben, C. (1977). *J. Mater. Sci.* **12,** 1325–1337.

Zweben, C. and Rosen, B. W. (1970). *J. Mech. Phys. Solids* **18,** 189–206.

Index